CYTOGENETICS - PAST, PRESENT AND FURTHER PERSPECTIVES

Edited by **Marcelo Larramendy**
and **Sonia Soloneski**

Cytogenetics - Past, Present and Further Perspectives
http://dx.doi.org/10.5772/intechopen.73451
Edited by Marcelo Larramendy and Sonia Soloneski

Contributors

Elena Salina, Irina Adonina, Aparecido Divino Da Cruz, Irene Plaza Pinto, Lysa Bernardes Minasi, Alex Silva Da Cruz, Emília Oliveira Alves Costa, Samara Socorro Silva Pereira, Fabilene Paim, Maria Lígia Nobile, Fausto Foresti, Claudio Oliveira, Aykut Yilmaz, Uliana Gorobeyko, Irina Kartavtseva, Natalia V. Kovaleva, Francesca Dumas, Rita Scardino, Vanessa Milioto, Mei-Chen Tseng

First published in London, United Kingdom, 2019 by IntechOpen
IntechOpen is the global imprint of INTECHOPEN LIMITED, registered in England and Wales, registration number: 11086078, The Shard, 25th floor, 32 London Bridge Street
London, SE19SG – United Kingdom
Printed in Croatia

British Library Cataloguing-in-Publication Data
A catalogue record for this book is available from the British Library

Additional hard copies can be obtained from orders@intechopen.com

Cytogenetics - Past, Present and Further Perspectives, Edited by Marcelo Larramendy and Sonia Soloneski
p. cm.
Print ISBN 978-1-83880-405-3
Online ISBN 978-1-83880-406-0
eBook (PDF) ISBN 978-1-83880-643-9

For EU product safety concerns :
IN TECH d.o.o., Prolaz Marije Krucifikse Kozulić 3, 51000 Rijeka, Croatia,
info@intechopen.com or visit our website at intechopen.com.

Meet the editors

Marcelo L. Larramendy, PhD, serves as a professor of Molecular Cell Biology at the School of Natural Sciences and Museum at the National University of La Plata (UNLP), Argentina. He is the senior researcher for the National Scientific and Technological Research Council of Argentina and a former member of the executive committee of the Latin American Association of Environmental Mutagenesis, Teratogenesis and Carcinogenesis. He has authored more than 450 contributions, including scientific publications, research communications, and conferences worldwide. He is a recipient of several national and international awards. Dr. Larramendy is a regular lecturer for Alexander Hollaender international courses organized by the IAEMS. He was a former guest scientist at the National Institutes of Health (NIH) in the United States and the University of Helsinki, Finland. He is an expert in genetic toxicology and has been a referee for more than 20 international scientific journals. He is a member of the International Panel of Experts at the International Agency for Research on Cancer (IARC, WHO, Lyon, France) for the evaluation of DDT, lindane, and 2,4-D. Currently, Dr. Larramendy is the head of the Laboratory of Molecular Cytogenetics and Genotoxicology at the UNLP.

Sonia Soloneski earned a PhD in Natural Sciences and is an assistant professor of Molecular Cell Biology at the School of Natural Sciences and Museum at the National University of La Plata (UNLP), Argentina. She is a member of the National Scientific and Technological Research Council (CONICET) of Argentina in the field of genetic toxicology. Currently, she is a member of the Latin American Association of Environmental Mutagenesis, Teratogenesis and Carcinogenesis (ALAMCTA), the Argentinean Society of Toxicology (ATA), and the Argentinean Society of Genetics (SAG). Dr. Soloneski has authored more than 350 scientific publications in the field, including scientific publications in research papers, reviewed journals, and conferences worldwide. She is a referent for issues related to genetic toxicology, mutagenesis, and ecotoxicology.

Contents

Preface

The classical definition of cytogenetics presented in Merriam-Webster, as employed since 1930, is "a branch of biology that deals with the study of heredity and variation by the methods of both cytology and genetics." The medical definition of cytogenetics includes the study of chromosomes, which are the visible carriers of heredity material. Taken together, these definitions convert cytogenetics into a fusion science joining cytology, the study of the cells, with genetics, the study of inherited variation.

Cytogenetics - Past, Present, and Further Perspectives discusses events that influenced the development of cytogenetics as a specialty within biology, with special attention paid to methodological achievements developed worldwide that have driven the field forward. Improvements to the resolution of chromosome analysis followed closely the introduction of innovative analytical technologies. In that sense, this book reviews and provides a brief account of the structure of chromosomes and stresses the high structural conservation in different species with an emphasis on aspects that require further research. However, it should be kept in mind that the future of cytogenetics will likely depend on improved knowledge of chromosome structure and function.

This book is organized into eight chapters and begins with an overview of cytogenetic methods employed to analyze homoeological chromosomes in cereals. There is special emphasis on the study of polyploid wheats and their progenitors, and on tandem repeats and retrotransposons as biomarkers to evaluate chromosome reorganization throughout the history of evolution and breeding. Chapter 2 provides a detailed karyotyping investigation of sixteen oak species belonging to three sections within the genera *Quercus* L. and includes a comparison between Turkish and European oaks. Chapter 3 describes several karyotypes and cytochrome b (cytb) gene sequences of seven species of *Epinephelus* grouper fish in order to facilitate future genetic breeding investigations. These include *Epinephelus coioides, E. flavocaeruleus, E. fuscoguttatus, E. lanceolatus, E. polyphekadion, E. tukula,* and *Plectropomus leopardus*. Chapter 4 focuses on the advantages and disadvantages of applying both morphological and phylogenetic tools in cytogenetic studies by investigating the structural-functional organization of macro- and microkaryotypes. This information helps one to better understand the genetic structures, evolution, and systematics of the richest and most diverse Neotropical fish groups. Chapter 5 presents a complete study of chromosomal characteristics in different common bat genera found in the Russian Far East and neighboring regions that improves the accuracy of the chromosome characteristics for 17 out of 18 valid species found in this region so far. Chapter 6 comprehensively analyses available global data on the epidemiology of Robertsonian translocations, the most common structural chromosomal rearrangements in humans. The information presented will allow future studies to resolve multiple unanswered questions, for example, the nature of female preponderance among carriers with Robertsonian translocations in newborns, the production of interchromosomal effects, and mosaicism. Chapter 7 reviews the evolutionary aspects of human chromosome 13 with classic cytogenetic methods, such as comparative banding, in combination with molecular cytogenetics techniques, such as chromosome painting and other in situ hybridization techniques. Overall, this chapter reconstructs the history of human chromosome 13 using a comparative approach across eutherian mammals. Finally, Chapter 8 describes the chromosomal microarray methodology that is employed for detecting and quantifying submicroscopic genomic gains and losses during DNA sample screening. The authors highlight that this methodology has valid applications in both diagnostic and functional scenarios, as well as potential applications in human genetic diagnosis, mutagenesis, agrigenomics, and pharmacogenomics, among other areas.

The editors of *Cytogenetics - Past, Present, and Further Perspectives* are enormously grateful to all colleagues and coworkers who have helped during the writing of this book for sharing their knowledge and insights. They have put extensive effort into gathering the information included in each chapter. We gratefully acknowledge the contributions made by the many specialists in this field of research.

We hope that the information presented in this book will meet the expectations and needs of all those interested in different aspects of cytogenetics, including scientists, physicians, pharmacologists, and students, among others. We especially hope that this book will guide those in the field to make new discoveries and employ different and previous investigations in their future research to understand both basic and applied aspects of cytogenetics.

Marcelo L. Larramendy, PhD and Sonia Soloneski, PhD
School of Natural Sciences and Museum
National University of La Plata
La Plata, Argentina

Chapter 1

Cytogenetics in the Study of Chromosomal Rearrangement during Wheat Evolution and Breeding

Elena A. Salina and Irina G. Adonina

Additional information is available at the end of the chapter

http://dx.doi.org/10.5772/intechopen.80486

Abstract

Cytogenetic methods such as chromosome banding and *in situ* hybridization remain relevant in the post-genomic era, especially for allopolyploid species where genome duplication in some cases makes it difficult to assess the reorganization of chromosomes during evolution. In this review, we give a brief description of cytogenetic methods for the analysis of homoeological chromosomes in cereals. Emphasis is placed on the development of methods for the study of polyploid wheat and its progenitors and on tandem repeats and retrotransposons as markers to evaluate chromosome reorganization throughout evolution and breeding. The most effective cytological probes used for the identification of chromosomes in wheat and Triticeae species by fluorescence and genomic *in situ* hybridization are described. Particular attention is paid to ribosomal genes used as markers in phylogenetic studies and for chromosome identification. Utility of these cytogenetic methods in the evaluation of breeding lines is demonstrated. A strategy for cytological analysis of wheat hybrids according to the degree of relationships between the species involved in crosses is also discussed.

Keywords: wheat, homoeological chromosomes, FISH, GISH, tandem repeats, retrotransposons, ribosomal genes

1. Introduction

The genus *Triticum* occupies a special position among cereals due to their different levels of ploidy and their adaptation to widely differing ecological and geographical regions of the world. The wealth of studies available on the origin of polyploid wheat provides the opportunity to comparatively analyze their genomes and those of putative donors, gaining insight into the reorganization of chromosomes in the process of evolution, domestication, and breeding.

Interest in chromosome organization and evolution in common wheat diploid progenitors and wild wheat is primarily because these species are a valuable source of new genes that were lost in the process of domestication.

The first allopolyploid of the Emmer wheat group is *Triticum dicoccoides* Koern. (2n = 28, genome BBAA), which arose as a result of hybridization of the diploid species *T. urartu* Thum. ex Gandil. and *Aegilops speltoides* Tausch (the most probable donor of the B-genome) [1]. Hexaploid wheat (*T. aestivum* L.) appeared about 7–10 thousand years ago as a result of a second round of hybridization between tetraploid wheat and the wild species *Ae. tauschii* Coss. (donor of the D-genome).

Wheat of Timopheevi group (tetraploids—*T. araraticum* Jakubz., *T. timopheevii* Zhuk., *T. militinae* Zhuk. et Migusch., and hexaploid *T. zhukovskyi* Menabde et Ericzjan) also came about through the hybridization of *T. urartu* and *Ae. speltoides*, but in another time scale. They have a genome designated as GGA^tA^t or GGA^tA^tDD, pointing at their partial homology with the genomes of wheat species in the Emmer group [2].

Diploid progenitors and species of the Timopheevi group are the source of new genes for the resistance to biotic and abiotic stresses. They are regularly involved in hybridization with common wheat, either directly or through synthetic polyploid forms. In this regard, knowledge on the structure and evolution of chromosomes of common wheat, its cultural and wild relatives is particularly relevant.

Genome sequencing methods actively displace cytogenetic analysis in current research. However, in working with hybrid material and introgressed lines of common wheat, we argue that the relevance of cytogenetic methods endures in the post-genomic era.

This review gives a brief description of the cytogenetic methods that remain relevant at this time and their use in the study of chromosomal rearrangement during wheat evolution and breeding.

2. Development of cytogenetic methods for studying chromosomes of polyploid wheat and their progenitors

2.1. Chromosome banding and *in situ* hybridization

Classical cytogenetic methods such as chromosome banding are currently relevant for wild species of plants as well as for polyploid species. Genome duplication within polyploids in some cases makes it difficult to assess the reorganization of chromosomes during evolution and hybridization.

The development of chromosome banding techniques allowed for the identification of the chromosomes not only in morphology but also in individual-specific patterns. Descriptions of the results of chromosome banding are based on the chromosomal region (band) and the intensity of staining, which differs from the neighboring regions. There are several methods of chromosome banding, namely, C-, N-, F-, Hy-, G-, Re-, and AgNOR-banding [3]. The most common method of staining used in the analysis of cereal genomes is C-differential staining (C-banding), first demonstrated by Pardu and Gal [4]. This method identifies the regions of constitutive heterochromatin after denaturation of the chromosomes and subsequent processing by Giemsa reagent. The karyotypes of many cereal crops, including polyploid wheat and their wild relatives were characterized based on C-banding [3, 5–8]. The use of C-banding allows the study of chromosomal rearrangement during evolution and breeding. For example, Badaeva with co-authors [9] used C-banding to analyze 460 samples of polyploid wheat and 39 forms of triticale (x *Triticosecale* Wittmack) from 37 countries. Fifty-eight main types of chromosomal rearrangements were identified. The results obtained by the authors showed that chromosomes of the B genome are more often involved in chromosomal rearrangements than chromosomes of A and D genomes.

Thus, it is clear that studies like these are necessary for a better understanding of the laws of evolutionary processes in the plant world. C-banding is also currently used to characterize hybrid material and wheat cultivars, especially when other methods of analysis do not reveal chromosome polymorphism [10].

In addition to the differential staining, a specific pattern on chromosomes can be obtained by hybridization *in situ*. Hybridization *in situ* is a direct method of localizing DNA sequences on chromosomes. It is based on the ability of denatured DNA molecules to form duplexes with homologous DNA sequences of chromosomes on a slide. *In situ* hybridization was first performed on animal chromosomes [11] and later applied to plants chromosomes [12]. Over its 50-year history, this method has undergone significant changes aimed at increasing the sensitivity in the detection of labeled probes. This is primarily due to the development of simpler and more efficient DNA tagging systems and better visualization of the hybridization signal. Currently, fluorescence *in situ* hybridization (FISH) is used to study the distribution of individual DNA sequences on chromosomes. Genomic *in situ* hybridization (GISH) is commonly used to identify alien DNA or to study the genomic composition of wheat amphiploids and hybrids.

2.2. Repetitive DNA as a source of markers for chromosome painting

Most often, various repetitive DNA sequences are used as probes for FISH. This is not surprising since repeats are the largest and most rapidly evolving part of the genome. According to the latest sequencing data, repetitive DNA accounts for about 80% of the cereal genome [13, 14]. Groups of repeats with similar structure, formed by amplification from a common original sequence, are called families. Families of repeats differ in their structure, the size of the monomer (from one to several thousand nucleotide pairs), the number of copies, and the type of proliferation.

Transposable elements are the most common repeat elements and account for more than 90% of the entire fraction of cereal repetitive DNA. All families of transposable elements are united into two larger categories—classes, according the mechanism of transposition (retrotransposons and DNA transposable elements). The current detail classification of transposable elements was described by Wicker et al. [15]. Mostly, the transposable elements are dispersed on chromosomes.

There are families of repeats whose members are organized in tandem and assembled into one or more loci. Depending on the length of the repeating unit (monomer), tandem repeats are divided into microsatellites (monomer length 1–6 bp), minisatellites (from 10 to 60 bp), and satellites (average monomer length from 100 to 700 bp) [16].

In fact, each chromosome has an individual "pattern" of repeats, which can be used effectively for marking and identification of individual chromosomes, and the whole genome.

2.2.1. Tandem repeats as markers to study the reorganization of chromosomes in the process of evolution

This group of repetitive DNA sequences is well studied in plants, especially cereals, and is widely used as markers in genomic research and in identifying chromosomes. According to their distribution on chromosomes, the repetitive sequences can be classified as centromeric, subtelomeric or intercalary. In combination, they generate a diagnostic "pattern" on the chromosome. Tandem repeats, such as microsatellites and satellites, and genes of ribosomal RNA are most frequently used for marking the chromosomes of wheat and its relatives.

Microsatellites are repeats with motifs from 1 to 6 bp. In plant genomes, they are also referred to as simple sequence repeats (SSRs) [17]. Microsatellites are used extensively as PCR markers for mapping chromosomes of many plant species and for gene labeling in applied research. Microsatellites are also used as cytogenetic markers. There are a few studies in which the distribution of various microsatellites on *T. aestivum* chromosomes has been examined in detail using FISH [18, 19]. For example, the dinucleotide probes $(AT)_{10}$ and $(GC)_{10}$ recorded no signal on chromosomes. This confirms the earlier hypothesis that the wheat genome does not contain extended clusters of these microsatellites [20, 21]. A dispersed distribution on chromosomes was established for probes $(AC)_8$ and $(GCC)_5$. The large microsatellite blocks detected by the probes $(AGG)_5$, $(CAC)_5$, $(ACG)_5$, $(AAT)_5$, and $(CAG)_5$ were found mainly in the pericentromeric regions of the B genome. Strong intercalary signals were detected after hybridization with the probe $(ACT)_5$ on a number of chromosomes of A and B genomes. Molnar and co-authors [22] investigated the distribution of microsatellites (ACG)n and (GAA) n on the chromosomes of *Ae. biuncialis* Vis (2n = 4x = 28, $U^bU^bM^bM^b$) and *Ae. geniculata* Roth. (2n = 4x = 28, $U^gU^gM^gM^g$) and on the chromosomes of their diploid progenitors: *Ae. umbellulata* Zhuk (UU) and *Ae. comosa* Sm. In Sibth.& Sm. (MM). They concluded that the break points for intergenomic translocations are often localized in regions saturated with microsatellite repeats. Thus, a number of studies have demonstrated that probes based on microsatellites can be useful for the identification of chromosomes and for a better understanding of the principles of chromatin organization in cereals. An important methodologically significant result was obtained by Cuadrado and Jouve [23]. They found that labeled oligonucleotides with a repeating mono-, di-, tri-, or tetra-nucleotide motif have the unexpected ability to detect the corresponding SSR loci even on nondenatured chromosomes, which in some cases can greatly facilitate and accelerate cytological analysis.

The (GAA)n microsatellite is the most widely used marker for the identification of chromosomes. The first works on its localization in the genome of cereals were carried out at the end of the twentieth century [24]. The GAA microsatellite was used as a marker for identification and sorting of polyploid wheat chromosomes [25, 26]. Phylogenetic studies using the GAA microsatellite were previously problematic due to the deficiency of hybridization signals on the A- and D-genomes and the presence of a number of major hybridization sites on the B-/G-genomes [19, 26].

In recent years, additional publications using GAA microsatellites for the identification of chromosomes of the A-genome of diploid wheat species and for phylogenetic analysis have appeared. Two works published in 2012 included data on the karyotypic analysis of single samples of *T. monococcum* L. and *T. urartu* using the oligonucleotide probe $(GAA)_9$ or GAA fragments obtained by PCR from genomic DNA of wheat [27, 28]. We performed a comparative analysis of the A-genome chromosomes in a diploid and polyploid wheat species consisting of two evolutionary lineages, Timopheevi and Emmer, using the pTm30 probe cloned from the *T. monococcum* genome and containing $(GAA)_{56}$ microsatellite sequences (**Figure 1**) [29].

Up to four pTm30 sites located on 1AS, 5AS, 2AS, and 4AL chromosomes have been revealed in the wild diploid species, although most accessions contained one to two (GAA)n sites (**Figure 1**). The (GAA)n loci on chromosomes 2AS, 4AL, and 5AL found in *T. dicoccoides* were retained in *T. durum* Desf. and *T. aestivum*. In species of the Timopheevi lineage, only one large (GAA)n site has been detected in the short arm of the $6A^t$ chromosome [29].

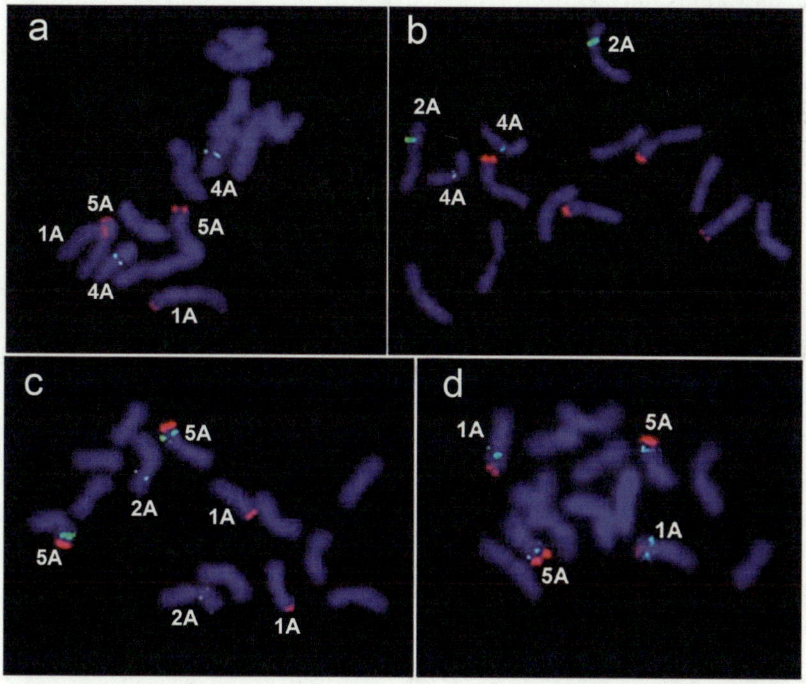

Figure 1. FISH with probes (GAA)n (green) and pTa71 (red) on the chromosomes of diploid *Triticum* species: (a) *T. monococcum* (acc. K-18140), (b) *T. boeoticum* Boiss (acc. K-25811), (c) *T. boeoticum* (acc. PI427328), and (d) *T. urartu* (acc. IG45298).

It was shown that changes in the distribution of (GAA)n sequences on the A-genome chromosomes of diploid and polyploid wheats are associated with chromosomal rearrangements/ modifications involving mainly the NOR (nucleolus organizer region)-bearing chromosomes, throughout the evolution of wild and domesticated species.

Satellite DNA repeating units are longer than 100 bp. These sequences are characterized by a high copy level (10^4–10^6) and form clusters of repeats, the length of which is rather difficult to estimate by high-throughput sequencing of genomes due to the "ejection" of the main part of tandem repeats during this process. Earlier studies of cereal genomes using pulsed field gel electrophoresis made it possible to estimate the length of the tandem repeat regions in a cluster as 90–600 kb [30]. Satellite DNA can comprise up to 5% of the genome and is the cause of significant differences in the content of heterochromatin DNA blocks in closely related species. Due to the high copy numbers of satellite DNA in the chromosomal locus, they are well detected in the FISH assay.

We can distinguish the following families of satellite DNA, whose units (in the form of cloned DNA sequences or PCR fragments) have been successfully used for the analysis of the genome of wheat and Triticeae species, including the study of the reorganization of genomes during evolution:

1. A family of repeats pAs1/Afa/pHcKB6/dpTa1 [31, 32], localized predominantly in the subtelomeric and intercalary chromosome regions of *Ae. tauschii*, D-genome of *T. aestivum*, species of the genus *Hordeum*, *Elymus*, and several other species.
2. The family of repeats 120 bp/pSc119.2 [33], widely distributed in subtelomeric and intercalary regions of chromosomes in many species of the tribe Triticeae (**Figure 2**) and in the closely related tribe Avenae [34]. This family of repeats was first isolated from *Secale cereale* L. and described as one of the families of telomeric rye heterochromatin [35].
3. The family of repeats 350 bp/pSc200/pSc74 and pSc250, which are the main tandem repeats of telomeric heterochromatin in rye *Secale cereale* [36]. During evolution, these sequences were amplified in the genome of individual species of *Secale*, as well as in certain species of the genera *Agropyron* and *Dasypyrum* of the Triticeae tribe.
4. pAesKB52/pGC1R-1/Spelt52 are tandem repeats of subtelomeric regions of chromosomes *Ae. speltoides*, *Ae. longissima* Schweinf, & Musch L., and *Ae. sharonensis* Eig (**Figure 2**) [37].
5. Spelt1 is a genome-specific sequence associated with telomeric heterochromatin of *Ae. speltoides* (**Figure 2**). Sequences of this family have not yet been detected by hybridization methods in the genomes of other Triticeae species, with the exception of *T. monococcum* (weak hybridization signal) and polyploid species formed with the participation of *Ae. speltoides* [37].

The probes, pSc119.2 and pAs1, are most often used for intraspecific identification of Triticeae tribe chromosomes by the FISH method. Thus, simultaneous hybridization of two DNA probes (pSc119.2 and pAs1) makes it possible to identify 17 (out of 21) chromosomes of the genome of common wheat [34, 38].

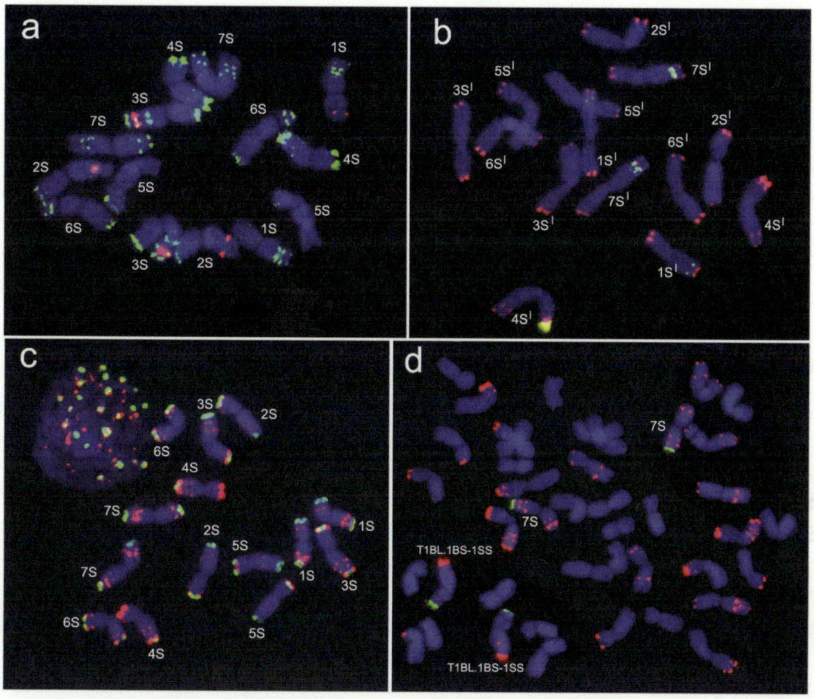

Figure 2. FISH to mitotic metaphase chromosomes. (a and c) *Ae. speltoides*, (b) *Ae. longissima*, and (d) hybrid line (*T. aestivum* × *Ae. speltoides*). Probe combinations used were (a) pSc119.2 (green) and Spelt52 (red), (b) pSc119.2 (red) and Spelt52 (green), and (c and d) pSc119.2 (red) and Spelt1 (green).

The combination of pSc119.2 and Spelt52 probes is effective for the study of all *Aegilops* species of the Sitopsis section (putative donors of the B/G genomes of polyploid wheats) (**Figure 2**). The probe combination pSc119.2 and Spelt1 is effective for only *Ae. speltoides* chromosome identification (**Figure 2**) [37]. The S-genomes of *Aegilops* species within the Sitopsis section are very similar to the common wheat B-genome. Accordingly, GISH with the DNA of these species is difficult on hybrid wheat lines. Therefore, we used FISH with the Spelt1 and Spelt52 probes to identify the *Aegilops* genetic material in the investigated lines (**Figure 2**). Simultaneous hybridization with probes pSc119.2 and pAs1 allows identification of wheat chromosomes.

The different level of homology within various families of tandem repeats depends on the rate of homogenization of repeats within the cluster, within each genome and species. The Spelt1-family is highly conserved, according to the sequencing of 10 Spelt1 sequences isolated from different accessions of *Ae. speltoides*, which shows a high level of homology (98%) [37]. It should be noted that more often a high level of interspecific polymorphism of satellite DNA families was demonstrated. This suggests that it may be possible to obtain efficient probes for the identification of chromosomes of the studied species by cloning DNA sequences of the repetitive families from these species.

Reports of the discovery of new families of highly repetitive DNA sequences are still emerging despite the existence of a large number of cereal tandem repeats already described. A previously unknown class of repeating DNA sequences named "Fat" was identified in the genome of common wheat [39]. Fat repeats are organized in clusters but with a dispersed distribution throughout the genome. The Fat-element content varies considerably across the genomes of different cereal species. The highest intensity of hybridization was found in the D-genome of wheat and *Aegilops* and in the S genome of *Agropyron*. This sequence was not found in oats or domesticated barley *Hordeum vulgare* L, but was present in minimal amounts in other species of the genus *Hordeum*. Based on this, it was concluded that the Fat-element first appeared in the evolution of cereals after the divergence of oats, during the separation of domestic barley from related grasses. The most intensive hybridization of Fat-repeats in the form of a large cluster of signals characterizes the chromosomes of the 4th homoeological group of wheat and *Aegilops*. This sequence is found only on the chromosomes of the D-genome of wheat and *Aegilops*, enabling it to be used as a FISH marker for identifying chromosomes and studying chromosome reorganization during evolution.

To discover new probes, 2000 plasmid wheat clones were examined by Komuro et al. [40]. Among them, 47 clones produced strong discrete signals on wheat chromosomes. Especially, valuable is combining pTa-535, pTa-713, and pTa-86 (pSc119 homolog) sequences, which allows to completely identify all 21 wheat chromosome pairs.

In addition, it seems promising to use oligonucleotides synthesized for various sites in the above-mentioned families of repeats identified in high-throughput sequencing, including sequencing data on individual chromosomes of wheat made for the identification of chromosomes. These probes have been shown to provide an easier, faster and more cost-effective method for the FISH analysis of wheat and hybrids [41, 42].

2.2.2. Ribosomal genes

An important and well-studied family of tandem repeats is the family of ribosomal RNA genes (rRNA). A detailed analysis of the monomers and cluster organization was carried out for these genes. Their localization on chromosomes in various species of cereals and possible mechanisms of evolutionary variability, including the processes of divergence and homogenization, were described by Hillis et al. [43]. There are two classes of rRNA-genes in the cereal genome: genes encoding 5S rRNA and 45S rRNA. 45S and 5S rDNA are located independently of each other, even in cases when they are both localized on one arm of a chromosome.

The 45S rDNA of cereals contains a coding region and a nontranscribed spacer sequence. The gene region includes three DNA sequences encoding 18S, 5.8S, and 26S rRNA, which are separated by internally transcribed spacer sequences. Polymorphism exists in the number of 45S rDNA loci in the Triticeae species genomes. The "major" loci of these genes are located on the short arms of homoeologous chromosome groups 1, 5, and 6. Nucleolus-forming regions are found on chromosomes 1A, 1B, 6B, and 5D of *T. aestivum* [44]. In addition, minor loci of 45S rDNA are also present, in which active RNA synthesis is not observed.

The genes encoding the 5S rRNA have the smallest repeating unit length among the ribosomal genes (320–500 bp). The repeating unit of 5S rDNA contains a 120 bp conserved coding region and a variable nontranscribed spacer sequence. There are from 1000 to 4000 copies of 5S rRNA genes per haploid genome in cereals. Two subfamilies of 5S rDNA are distinguished in the Triticeae genomes, depending on the length of the spacer: 5SDna1 (200–345 bp) and 5SDna2 (350–380 bp). Hybridization of 5S-repeats on chromosomes of various *Triticeae* species showed that in most species, they are located in homoeological groups 1 and 5. An analysis of the chromosome distribution of 5SDna1 and 5SDna2 subfamilies showed that the short units of 5S rDNA have preferential localization on the chromosomes of homoeologous group 1, while the long units are located on group 5. It was shown that 5S rRNA genes with a monomer length of 290 bp are located on chromosomes 1B and 1D of common wheat, and genes with a monomer length of 410 bp are located on chromosomes 5B and 5D [45]. Further work on the isolation and sequencing of individual monomers led to division of the 5S rRNA genes into a larger number of subfamilies [46].

Analysis of chromosome 5B sequencing data, as well as individual BAC-clones containing 5S rDNA, showed that long and short types of subunits can be located on one chromosome, but they form separate clusters interrupted by the insertion of mobile elements [47].

The presence of conservative (coding) and polymorphic (noncoding) sequences in rDNA promoted their widespread use as molecular markers in phylogenetics. 5S and 45S rDNA are also widely used as cytogenetic markers for FISH due to their large copy number and localization in certain regions of chromosomes. A number of phylogenetic studies using individually cloned copies of 45S and 5S rDNA have been carried out for wheat and its relatives [48, 49]. An interesting fact is that among *Triticum* and *Aegilops*, two species (*T. timopheevii* and *Ae. speltoides*) lost the 5S rDNA locus on the chromosome of homoeologous group 1 (1G and 1S, correspondently) during evolution [50].

2.2.3. *Transposable genetic elements as markers of genomic rearrangements*

Another class of repetitive DNA, widely represented in the genome of plants, is transposable genetic elements (TEs), which are divided into two classes: class I elements (retrotransposons) and class II elements (DNA transposable elements).

At present, it seems likely that the diverse TEs, which have a mainly dispersed chromosomal localization, are the major contributors to the observed interspecies differentiation of chromosomes revealed by genomic *in situ* hybridization (GISH). GISH, a method based on the hybridization of labeled genomic DNA of one species to metaphase chromosomes of another species or hybrid, is widely used to assess the degree of genome homology. GISH serves as a unique approach to studying the formation of genomes of polyploid species and revealing the nature of their relationship, the analysis of introgression of alien genetic material, and the localization of break points in intergenomic translocations in remote hybrids [51, 52].

The development of BAC (bacterial artificial chromosome) libraries containing clones with very large inserts (>100 kb) of genomic DNA has opened up new possibilities for studying the reorganization of genomes by BAC *in situ* hybridization (BAC-FISH). The localization of BAC clones on chromosomes is mainly connected with families of TEs in their composition. Thus, carrying out BAC-FISH on wheat chromosomes showed a different BAC localization in the genome depending on which family of TEs or other repeats were present in them [53, 54].

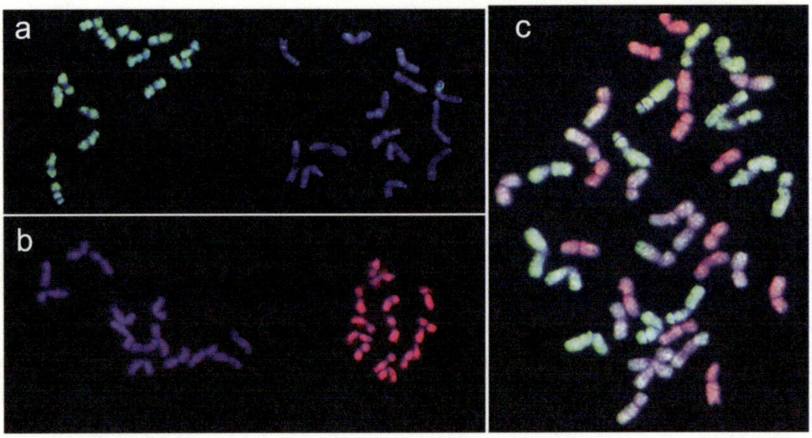

Figure 3. FISH to mitotic metaphase chromosomes of (a) *Ae. speltoides* and *T. urartu*, (b) *T. urartu* and *Ae. tauschii*, and (c) *T. aestivum*. With the probe combinations: (a) BAC clone 2383A24 (green), (b) BAC clone 112D20 (red), (c) BAC clone 2383A24 (green), and BAC clone 112D20 (red).

It is interesting to note the differential amplification of individual members of retrotransposon families belonging to the subclass Ty3-*gypsy*-retrotransposons in the genomes of diploid species, which retains genomic specificity in allopolyploid wheats (**Figure 3**) [55, 56].

Thus, FISH analysis of a BAC clone (BAC_2383A24) demonstrated its predominant localization to chromosomes of the B-genome of allopolyploid wheats and its putative diploid progenitor *Ae. speltoides* (**Figure 3**).

Analysis of the complete BAC_2383A24 nucleotide sequence revealed that three elements of the gypsy LTR retrotransposon family *Fatima* make up 47.2% of all the LTR retrotransposons in this BAC. Phylogenetic analysis, as well as FISH, showed that these *Fatima* elements are predominantly from the B genome of common wheat and its putative progenitor *Ae. speltoides* (**Figure 3**). Similar approaches, including hybridization with BAC clone 112D20, demonstrated that the *Lila* family of Ty3-*gypsy*-retrotransposons is predominantly from the D-genome and its progenitor *Ae. tauschii*. Multiple FISH with both clones allows the identification of all three subgenomes of hexaploid wheat (**Figure 3**).

Dating of the LTR retrotransposon insertion showed that TE proliferation mainly occurred in this diploid species before it entered into allopolyploidy [55, 57].

3. Chromosomal rearrangement during wheat breeding

Genetic erosion, caused by modern agricultural breeding practices, has led to the observed decrease in genetic variation in crops, including common wheat *T. aestivum*. Wheat relatives—wild and cultivated cereals—are used as sources of effective genes for resistance to biotic and abiotic stresses and to increase genetic diversity.

Introgression of genes from related species to wheat depends on the level of divergence between the species involved in the cross. Species belonging to the primary gene pool have homoeologous genomes. This group includes wild and cultivated forms of *T. turgidum* and species of donors A and D of the genomes of common wheat: *T. urartu*, *T. monococcum*, and *T. boeoticum*, as well as *Ae. tauschii*. The transfer of genes from these species can be carried out by crossing, homoeologous recombination, backcrossing and selection, as well as through the development of synthetic amphiploids.

The secondary gene pool includes polyploid species of wheat and *Aegilops* which have at least one homoeologous genome with *T. aestivum*. The transfer of genes from these species to common wheat by means of homoeologous recombination is also possible if recombination has taken place between the target homoeologous chromosomes. This group also includes hexaploid species with GGA^tA^tDD genome: *T. kiharae*, Dorof. et Migusch., *T. miguschovae*; tetraploid species with GGA^tA^t genome: *T. timopheevii*, *T. militinae*, and *T. araraticum*; and diploid species of *Aegilops* from the Sitopsis section, which are close to the B genome of *T. aestivum*. Cytogenetic analysis of hybrids from crosses of common wheat with *T. timopheevii* showed homoeologous introgression of G genome fragments to practically all chromosomes of both the B genome and the D genome of common wheat [58]. It should also be noted that the extent of introgressive regions varies among wheat lines [59]. Genetic material from *Ae. speltoides* (SS genome), the putative progenitor of the B and G genomes of polyploid wheat, was successfully transferred to all three genomes of common wheat, but especially, as expected, in the chromosome of the B-genome [60, 61].

Species that do not carry the genomes A, B, and D, and those related to the tertiary gene pool, are considered more distant relatives of wheat. The transfer of genes from these species is difficult since it cannot be accomplished by recombination and therefore requires the use of other strategies. Currently, there are standard methods that facilitate the transfer of genes from species that do not have related genomes with common wheat. Some are based on the methods of chromosome engineering, and others manipulate the genetic control of meiotic recombination or employ genetic engineering. The transfer of genetic material in this case occurs both in the partly homoeologous group of chromosomes and into other groups [62].

The strategy used in cytological analysis of hybrids depends first on the nature of the relationships between the species involved in crossing. In instances where the donor species belongs to the tertiary gene pool with respect to *T. aestivum*, GISH is first used, which allows the estimation of the size and localization of the alien translocation. GISH can be used successfully to identify translocations of rye, wheatgrass, and *Aegilops* species (with the exception of the Sitopsis group) in the wheat genome (**Figure 4**).

However, GISH does not answer the questions: which wheat chromosome is replaced by an alien chromosome or which alien chromosome took part in the translocation. In addition, if the genomes of the crossed species are evolutionarily close, that is, if donor species refer to primary and secondary gene pools, then GISH will also be difficult. A similar problem occurs, for example, in the analysis of hybrids from the crossing of hexaploid wheat with *Ae.*

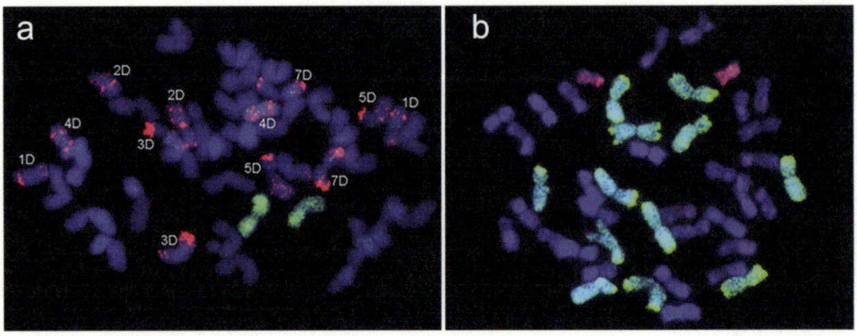

Figure 4. GISH with *Th. intermedium* DNA (green) in combination with pAs1 probes (red) of common wheat cultivar Tulaikovskaya 5 (a) and GISH with *S. cereale* DNA (green) and with *Ae. umbellulata* DNA (red) of the triticale line with introgression from *Ae. umbellulata* (b).

speltoides (B-genome putative donor) or with the species *Ae. tauschii* which is a donor of the D genome. In this case, species-specific markers are used to identify chromosomes in hybrids or introgressive wheat lines, such as, for example, Spelt1 for the *Ae. speltoides* genome (**Figure 2**).

It should be noted that when identifying the alien introgressions in lines/varieties developed by remote hybridization of cereals, best results are usually obtained by combining different methodological approaches and using different chromosomal markers.

4. Conclusion

The wheat allopolyploids have long attracted the attention of researchers, both from the perspective of studying the processes of genome reorganization during amphiploidization and to develop new wheat lines for breeding.

To accomplish these fundamental and applied tasks, various approaches are used. In recent years, SNP markers and various technologies for their identification have been actively involved, as well as reference genome data for wheat and related species. This allows us to obtain more detailed information about the organization and evolution of the wheat genome and the structure of gene families present in reference genomes. Despite continued progress in deciphering the complex wheat genome, a complete understanding of the reorganization of the wheat genome during evolution can only be obtained by combining molecular methods of analysis with cytogenetic ones. The latter makes it possible to identify rearrangements of homoeological chromosomes in the process of evolution and breeding.

The first translocations in wheat varieties were detected by cytogenetic methods. Later, the molecular markers developed for these translocations allowed the use of marker-assisted breeding for selection of the desired genotypes.

One of the most successful used in selection is the translocation of the short 1R chromosomal arm to 1A and 1B of the wheat chromosome during breeding. At present, more than 300 soft wheat varieties carry the T1RS.1BL translocation [http://www.rye-gene-map.de/rye-introgression], which determines the resistance to phytopathogens and increased productivity. Interestingly, the presence of an intact wheatgrass chromosome in Russian wheat varieties was found to be significant for resistance to fungal diseases and the maintenance of grain quality over the last 30 years [10]. It should be noted that only a set of C-banding methods used in conjunction with *in situ* hybridization and assays with PLUG and SSR markers revealed that wheat chromosome 6D in the wheat cultivars was substituted by the *Thinopyrum intermedium* (Host) homoeologous chromosome, 6Ai.

It should be emphasized once again that, despite extensive development of molecular markers for genome analysis, including high-throughput genotyping, it is impossible to characterize the modern diversity within the genus *Triticum* without involving cytogenetic methods.

Acknowledgements

We are grateful to Carly Schramm for critical review of the manuscript. This work was supported by the Russian Foundation for Basic Research, project no. 17-04-00507 and the IC&G Budgetary project no. 0324-2018-0018.

Conflict of interest

The authors declare that they have no competing interests.

Author details

Elena A. Salina* and Irina G. Adonina

*Address all correspondence to: salina@bionet.nsc.ru

The Federal Research Center Institute of Cytology and Genetics of Siberian Branch of the Russian Academy of Sciences, Russia

References

[1] Feldman M. The origin of cultivated wheat. In: Bonjean A, Angus W, editors. The Wheat Book. Paris: Lavoisier Publishing; 2001. pp. 3-58

[2] Mitrofanova O, Badaeva E, Salina EA. *T. timopheevii* and *T. zhukovskii*, the bread wheat cousins and their contribution to *T. aestivum* improvement. In: Bonjean AP, Angus WJ, van Ginkel M, editors. The World Wheat Book: A History of Wheat Breeding; part 4, chapter 39. Vol. 3. France: Tec & Doc Lavoisier; 2016. pp. 1167-1228

[3] Friebe B, Gill BS. Chromosome banding and genome analysis in diploid and cultivated polyploid wheats. In: Jauhar PP, editor. Methods in Genome Analysis in Plants. Boca Raton: CRC Press; 1996. pp. 39-60

[4] Pardue ML, Gall JG. Chromosomal localization of mouse satellite DNA. Science. 1970; **168**:1356-1358

[5] van Niekerk HA, Pienaar R. Morphology and linear C-band differentiation of *Triticum aestivum* cv. "Chinese Spring" chromosomes. Cereal Research Communications. 1983; **11**:115-122

[6] Gill BS, Friebe B, Endo TR. Standard karyotype and nomenclature system for description of chromosome bands and structural aberrations in wheat (*Triticum aestivum*). Genome. 1991;**34**:830-839. DOI: 10.1139/g91-128

[7] Badaeva ED, Keilwagen J, Knüpffer H, Waßermann L, Dedkova OS, Mitrofanova OP, et al. Chromosomal passports provide new insights into diffusion of Emmer wheat. PLoS One. 2015;**10**(5):e0128556. DOI: 10.1371/journal.pone.0128556

[8] Badaeva ED, Shishkina AA, Goncharov NP, Zuev EV, Lysenko NS, Mitrofanova OP, et al. Evolution of *Triticum aethiopicum* Jakubz. from the position of chromosome analysis. Russian Journal of Genetics. 2018;**54**(6):629-642. DOI: 10.1134/S1022795418060029

[9] Badaeva ED, Dedkova OS, Gay G, Pukhalskyi VA, Zelenin AV, Bernard S, et al. Chromosomal rearrangements in wheat: Their types and distribution. Genome. 2007;**50**:907-926. DOI: 10.1139/g07-072

[10] Salina EA, Adonina IG, Badaeva ED, Kroupin PY, Stasyuk AI, Leonova IN, et al. A thinopyrum intermedium chromosome in bread wheat cultivars as a source of genes conferring resistance to fungal diseases. Euphytica. 2015;**204**:91-101. DOI: 10.1007/s10681-014-1344-5

[11] Gall JG, Pardue ML. Formation and detection of RNA-DNA hybrid molecules in cytological preparations. Proceedings of the National Academy of Sciences of the United States of America. 1969;**63**:378-383

[12] Brady T, Clutter ME. Cytolocalization of ribosomal cistrons in plant polytene chromosomes. The Journal of Cell Biology. 1972;**53**(3):827-832. DOI: 10.1083/jcb.53.3.827

[13] Mayer KFX, Rogers J, el Dole J, Pozniak C, Eversole K, Feuillet C, et al. A chromosome-based draft sequence of the hexaploid bread wheat (*Triticum aestivum*) genome. Science. 2014;**345**(6194):1251788. DOI: 10.1126/science.1251788

[14] Ling HQ, Ma B, Shi X, Liu H, Dong L, Sun H, et al. Genome sequence of the progenitor of wheat A subgenome *Triticum urartu*. Nature. 2018;**557**(7705):424-428. DOI: 10.1038/s41586-018-0108-0

[15] Wicker T, Sabot F, Hua-Van A, et al. A unified classification system for eukaryotic transposable elements. Nature Reviews. Genetics. 2007;**8**(12):973-982. DOI: 10.1038/nrg2165

[16] Salina EA. Tandem repeats in evolution of polyploid wheat and Aegilops section Sitopsis. Israel Journal of Plant Sciences. 2007;**55**(3-4):231-240. DOI: 10.1560/IJPS.55.3-4.231

[17] Morgante M, Olivieri AM. PCR—Amplified microsatellites as markers in plant genetics. The Plant Journal. 1993;**3**(1):175-182. DOI: 10.1046/j.1365-313X.1993.t01-9-00999.x

[18] Cuadrado A, Cardoso M, Jouve N. Physical organization of simple sequence repeats (SSR) in *Triticeae*: Structural, functional and evolutionary implications. Cytogenetic and Genome Research. 2008;**120**:210-219. DOI: 10.1159/000121069

[19] Cuadrado A, Cardoso M, Jouve N. Increasing the physical markers of wheat chromosomes using SSRs as FISH probes. Genome. 2008;**51**:809-815. DOI: 10.1139/G08-065

[20] Kantety RV, La Rota M, Matthews DE, Sorrells ME. Data mining for simple sequence repeats in expressed sequence tags from barley, maize, rice, sorghum and wheat. Plant Molecular Biology. 2002;**48**:501-510

[21] La Rota M, Kantety RV, Yu J-K, Sorrells ME. Nonrandom distribution and frequencies of genomic and EST-derived microsatellite markers in rice, wheat, and barley. BMC Genomics. 2005;**6**:23. DOI: 10.1186/1471-2164-6-23

[22] Molnar I, Cifuentes M, Schneider A, Benavente E, Molnar-Lang M. Association between simple sequence repeat-rich chromosome regions and intergenomic translocation breakpoints in natural populations of allopolyploid wild wheats. Annals of Botany. 2010;**107**:65-76. DOI: 10.1093/aob/mcq215

[23] Cuadrado A, Jouve N. Chromosomal detection of simple sequence repeats (SSRs) using nondenaturing FISH (ND-FISH). Chromosoma. 2010;**119**:495-503. DOI: 10.1007/s00412-010-0273-x

[24] Pedersen C, Rasmussen SK, Linde-Laursen I. Genome and chromosome identification in cultivated barley and related species of the *Triticeae* (Poaceae) by in situ hybridization with the GAA-satellite sequence. Genome. 1996;**39**(1):93-104

[25] Vrána J, Kubaláková M, Simkova H, et al. Flowsorting of mitotic chromosomes in common wheat (*Triticum aestivum* L.). Genetics. 2000;**156**(4):2033-2041

[26] Kubaláková M, Kovářová P, Suchánková P, Číhalíková J, Bartoš J, Lucretti S, et al. Chromosome sorting in tetraploid wheat and its potential for genome analysis. Genetics. 2005;**170**:823-829

[27] Danilova TV, Friebe B, Gill BS. Single-copy gene fluorescence *in situ* hybridization and genome analysis: Acc-2 loci mark evolutionary chromosomal rearrangements in wheat. Chromosoma. 2012;**121**:597-611. DOI: 10.1007/s00412-012-0384-7

[28] Megyeri M, Farkas A, Varga M, Kovacs G, Molnar-Lang M, Molnar I. Karyotypic analysis of *Triticum monococcum* using standard repetitive DNA probes and simple sequence repeats. Acta Agronomica Hungarica. 2012;**60**(2):87-95. DOI: 10.1556/AAgr.60.2012.2.1

[29] Adonina IG, Goncharov NP, Badaeva ED, Sergeeva EM, Petrash NV, Salina EA. (GAA)n microsatellite as an indicator of the A genome reorganization during wheat evolution and domestication. Comparative Cytogenetics. 2015;**9**(4):533-547. DOI: 10.3897/CompCytogen.v9i4.5120

[30] Alkhimova OG, Mazurok NA, Potapova TA, Zakian SM, Heslop-Harrison JS, Vershinin AV. Diverse patterns of the tandem repeats organization in rye chromosomes. Chromosoma. 2004;**113**(1):42-52. DOI: 10.1007/s00412-004-0294-4

[31] Rayburn AL, Gill BS. Isolation of a D-genome specific repeated DNA sequence from *Aegilops squarrosa*. Plant Molecular Biology Reporter. 1986;**4**(2):102-109

[32] Vershinin AV, Svitashev S, Gummesson PO, Salomon B, Bothmer R, Bryngelsson T. Characterization of a family of tandemly repeated DNA sequences in *Triticeae*. Theoretical and Applied Genetics. 1994;**89**:217-225. DOI: 10.1007/BF00225145

[33] McIntyre CL, Pereira S, Moran LB, Appels R. New Secale sereale (rye) DNA derivatives for the detection of rye chromosome segments in wheat. Genome. 1990;**33**:317-323

[34] Mukai Y, Nakahara Y, Yamamoto M. Simultaneous discrimination of the three genomes in hexaploid wheat by multicolor fluorescence in situ hybridization using total genomic and highly repeated DNA probes. Genome. 1993;**36**:489-494

[35] Bedbrook JR, Jones J, O'Dell M, Thompson RD, Flavell RB. A molecular description of telomeric heterochromatin in Secale species. Cell. 1980;**19**:545-560

[36] Vershinin AV, Schwarzacher T, Heslop-Harrison JS. The large-scale organization of repetitive DNA families at the telomeres of rye chromosomes. The Plant Cell. 1995;**7**:1823-1833. DOI: 10.1105/tpc.7.11.1823

[37] Salina EA, Lim YK, Badaeva ED, Scherban AB, Adonina IG, Amosova AV, et al. Phylogenetic reconstruction of Aegilops section Sitopsis and the evolution of tandem repeats in the diploids and derived wheat polyploids. Genome. 2006;**49**:1023-1035. DOI: 10.1139/g06-050

[38] Schneider A, Linc G, Molnar-Lang M. Fluorescence in situ hybridization polymorphism using two repetitive DNA clones in different cultivars of wheat. Plant Breeding. 2003;**122**:396-400. DOI: 10.1046/j.1439-0523.2003.00891.x

[39] Badaeva ED, Zoshchuk SA, Paux E, Gay G, Zoshchuk NV, Roger D, et al. Fat element—A new marker for chromosome and genome analysis in the *Triticeae*. Chromosome Research. 2010;**18**:697-709. DOI: 10.1007/s10577-010-9151-x

[40] Komuro S, Endo R, Shikata K, Kato A. Genomic and chromosomal distribution patterns of various repeated DNA sequences in wheat revealed by a fluorescence *in situ* hybridization procedure. Genome. 2013;**56**(3):131-137. DOI: 10.1139/gen-2013-0003

[41] Fu SL, Chen L, Wang YY, Li M, Yang ZJ, Qiu L, et al. Oligonucleotide probes for ND-FISH analysis to identify rye and wheat chromosomes. Scientific Reports. 2015;**5**:10552. DOI: 10.1038/srep10552

[42] Xiao Z, Tang S, Qiu L, Tang ZX, Fu S. Oligonucleotides and ND-FISH displaying different arrangements of tandem repeats and identification of *Dasypyrum villosum* chromosomes in wheat backgrounds. Molecules. 2017;**22**:973. DOI: 10.3390/molecules22060973

[43] Hillis DM, Moritz C, Porter CA, Baker R. Evidence for biased gene conversion in concerted evolution of ribosomal DNA. Science. 1991;**251**:308-310

[44] Dubcovsky J, Dvorak J. Ribosomal RNA multigene loci: Nomads of the *Triticeae* genomes. Genetics. 1995;**140**:1367-1377

[45] Appels R, Baum BR, Clarke BC. The 5S rDNA unit of bread wheat *Triticum aestivum*. Plant Systematics and Evolution. 1992;**183**:183-194

[46] Baum BR, Bailey LG, Belyayev A, Raskina O, Nevo E. The utility of the non-transcribed spacer of the 5S rDNA units grouped into unit classes assigned to haplomes test on cultivated wheat and wheat progenitors. Genome. 2004;**47**:590-599. DOI: 10.1139/g03-146

[47] Sergeeva EM, Shcherban AB, Adonina IG, Nesterov MA, Beletsky AV, Rakitin AL, et al. Fine organization of genomic regions tagged to the 5S rDNA locus of the bread wheat 5B chromosome. BMC Plant Biology. 2017;**17**(Suppl1):143-155. DOI: 10.1186/s12870-017-1120-5

[48] Allaby RG, Brown TA. Network analysis provides insights into evolution of 5S rDNA arrays in Triticum and Aegilops. Genetics. 2001;**157**(3):1331-1341

[49] Mahelka V, Kopecky D, Baum BR. Contrasting patterns of evolution of 45S and 5S rDNA families uncover new aspects in the genome constitution of the agronomically important grass *Thinopyrum intermedium* (*Triticeae*). Molecular Biology and Evolution. 2013;**30**(9):2065-2086. DOI: 10.1093/molbev/mst106

[50] Badaeva ED. Chromosome analysis in investigation of the origin of the B- and G-genomes of polyploid wheats. Membrane & Cell Biology. 2001;**18**(3):216-229 (in Russian)

[51] Jiang J, Gill BS. Nonisotopic *in situ* hybridization and plant genome mapping: The first 10 years. Genome. 1994;**37**(5):717-725

[52] Belyaev A, Raskina O. Heterochromatin discrimination in *Aegilops speltoides* by simultaneous genomic *in situ* hybridization. Chromosome Research. 1998;**6**(7):559-565

[53] Zhang P, Li W, Fellers J, Friebe B, Gill BS. BAC-FISH in wheat identifies chromosome landmarks consisting of different types of transposable elements. Chromosoma. 2004;**112**:288-299. DOI: 10.1007/s00412-004-0273-9

[54] Salina EA, Sergeeva EM, Adonina IG, Shcherban AB, Afonnikov DA, Belcram H, et al. Isolation and sequence analysis of the wheat B genome subtelomeric DNA. BMC Genomics. 2009;**10**:414. DOI: 10.1186/1471-2164-10-414

[55] Salina EA, Sergeeva EM, Adonina IG, Shcherban AB, Belcram H, Huneau C, et al. The impact of Ty3-gypsy group LTR retrotransposons Fatima on B-genome specificity of polyploid wheats. BMC Plant Biology. 2011;**11**:99. DOI: 10.1186/1471-2229-11-99

[56] Shcherban' AB, Adonina IG, Salina EA. Impact of Ty3/Gypsy group retrotransposon Lila on the D-genome specificity of common wheat *Triticum aestivum* L. Molecular Biology. 2012;**46**(4):522-530. DOI: 10.1134/S002689331202015X

[57] Charles M, Belcram H, Just J, Huneau C, Viollet A, Couloux A, et al. Dynamics and differential proliferation of transposable elements during the evolution of the B and A genomes of wheat. Genetics. 2008;**180**:1071-1086. DOI: 10.1534/genetics.108.092304

[58] Badaeva ED, Budashkina EB, Bilinskaya EN, Pukhalskiy VA. Intergenomic chromosome substitutions in wheat interspecific hybrids and their use in the development of a genetic nomenclature of *Triticum timopheevii* chromosomes. Russian Journal of Genetics. 2010;**46**(7):769-785. DOI: 10.1134/S102279541007001X

[59] Timonova EM, Leonova IN, Röder MS, Salina EA. Marker-assisted development and characterization of a set of *Triticum aestivum* lines carrying different introgressions from the *T. timopheevii* genome. Molecular Breeding. 2013;**31**:123-136. DOI: 10.1007/s11032-0129776-x

[60] Liu W, Koo D-H, Friebe B, Gill B. A set of *Triticum aestivum-Aegilops speltoides* Robertsonian translocation lines. Theoretical and Applied Genetics. 2016;**129**:2359-2368. DOI: 10.1007/s00122-016-2774-3

[61] Adonina IG, Petrash NV, Timonova EM, Khristov YA, Salina EA. Construction and study of leaf rust_resistant common wheat lines with translocations of *Aegilops speltoides* Tausch. genetic material. Russian Journal of Genetics. 2012;**48**(4):404-409. DOI: 10.1134/S1022795412020020

[62] Friebe B, Yiang J, Raupp WJ, McIntosh A, Gill BS. Characterization of wheat-alien translocations conferring resistance to diseases and pests: Current status. Euphytica. 1996; **91**:59-87

Chapter 2

Cytogenetic Relationships of Turkish Oaks

Aykut Yılmaz

Additional information is available at the end of the chapter

http://dx.doi.org/10.5772/intechopen.78265

Abstract

The genus *Quercus* L. represented by 18 species belonging to three sections (*Quercus* L., *Ilex* Loudon and *Cerris* Loudon) in Turkey has a problematic taxonomy especially because of widespread hybridization observed frequently among oak species. In the present chapter, detailed karyotypic investigation of the 16 oak species from three sections in Turkey, their taxonomic relationships with each other and European oaks and finally situations that increase the variation and taxonomic problems among oak species in Turkey were evaluated to contribute to the taxonomic relationships of the genus. The somatic chromosome number in all taxa examined confirmed that the chromosome number of the genus is consistent with 2n = 24. However, some differences were determined among species, although similar results were observed in many oak species in Turkey contrary to European oaks.

Keywords: *Quercus*, *Ilex*, *Cerris*, Turkey

1. Introduction

The genus *Quercus* L. belonging to the Fagaceae family is one of the most abundant and economically important genera of woody plants in temperate and tropical areas of the northern hemisphere. The genus contains over 500 species and is distributed across the North and Central America, Asia, Europe and North Africa [1].

Turkey has a large land area covered by forest with 21.2 million ha. The oaks in Turkey are ecologically and economically very important trees and make up 26% of the total forest area. However, a large part of this is not so productive due to especially excessive cutting and grazing. Turkey is one of the richest countries in species number, diversity and endemic species [2]. The genus is currently represented by 23 taxa and 3 endemic species such as *Q. aucheri*, *Q. vulcanica* and *Q. macranthera* subsp. *syspirensis* [2, 3].

The most important reason which influences species number and diversity is the location and geomorphological structure of Turkey [4]. Turkey is composed of Anatolian and Thrace peninsulas where three phytogeographic regions are crossed such as Irano-Turanian, Mediterranean and Euro-Siberian regions. Its geomorphological structure reveals numerous climatic regions which influence the flora. The main climatic changes in Turkey occured especially at the beginning of the Holocene, and these climatic conditions have affected the topography and distribution of today's plant vegetation [5]. Today's plant zones occured in Anatolia, about 8000 years ago [5–11]. During the last glacial period, coastal plants, such as *Pinus sylvestris* and Picea sp., which were grown in cold climate in East Black Sea, have been taken up to higher places up gradually, and their places have been occupied by broad-leaved forests, such as *Quercus* spp., *Fagus*, and so on, living in the shaded coastal areas [5].

Another important factor affecting species diversity and distribution is the location of Turkey between the Asian and European continents. As a result, Anatolia has served as a migration route and refuge regions for many plants and animals from Asia into Southeast Europe [12].

Besides its location and phytogeographical regions, Anatolian Diagonal which divides Anatolia as eastern and western parts is another factor affecting species diversity, number and distribution [4, 12–15]. Anatolian Diagonal separates many plant and animal species into eastern and western Anatolia.

Turkey is a rich country in the aspect of oak variation with 18 species belonging to three sections (*Quercus* L., *Ilex* Loudon, and *Cerris* Loudon).

Section *Quercus* L. is characterized by the greatest number of species among the three sections in Turkey: *Q. pontica* C. Koch., *Q. robur* L., *Q. hartwissiana* Steven., *Q. macranthera* subsp. *syspirensis* (C. Koch.) Menitsky, *Q. frainetto* Ten., *Q. petraea* (Mattuschka) Lieb., *Q. vulcanica* (Boiss. Heldr. ex) Kotschy, *Q. infectoria* Oliver, *Q. pubescens* Willd and *Q. virgiliana* Ten. [2].

Section *Ilex* Loudon is characterized by the evergreen trees and shrubs: *Q. ilex* L., *Q. coccifera* L. and *Q. aucheri* Jaub. et Spach. [2].

Section *Cerris* Loudon is the second largest section and includes five species; *Q. libani* Olivier, *Q. trojana* Webb, *Q. cerris* L., *Q. brantii* Lindl. and *Q. ithaburensis* subsp. *macrolepis* (Kotschy) Hedge et Yalt. [2].

Vegetative characters are preferred instead of reproductive characters in situations which are not helpful in the classification [16]. Leaves are good indicators of putative hybridization and very useful for identification of oaks. Individuals that exhibit intermediate morphological characters can be seen widely because of interspecific hybridization and even sometimes exhibit high morphological variation, and it is not possible to identify an oak tree to a species. In this case, acorns are secondary important materials in the seperation of oak species [17–19]. Although vegetative characters are crucial to differentiating species within the genus and are freguently prefered in the identification of oaks, these are insufficient and risky because of hybridization behaviors caused by weak reproductive barriers between oak species. Oaks are wind-pollinated species and they can spread across wide geographic regions [20–22]. As a result, many oak species grow in mixed populations that increase the hybridization in the same or different sections. In addition to wind-pollination and weak reproductive barriers between species [14, 22–25], insufficient diagnostic morphological characters [24–25] and the lack of investigations for each taxon such as ecological, historical and genetic descriptors [25] make problematic the genus *Quercus* in Turkey and similarly in the world.

Scientific interest has recently moved from classification of the species with classic descriptors to understanding of oak evolution with molecular markers [24–27]. However, oak taxonomy is still problematic and under debate. Although cytologic studies are very important and useful for determinig taxonomy and polyploidy, studies on the genus *Quercus* are still insufficient because of the small size of chromosomes and difficulties in the germination of acorns.

2. Materials and methods for karyotype analyses of Turkish oaks

In our previous studies, acorns belonging to 16 oak species from three different sections were collected in different times from various locations in Turkey. The species, sections and locations are presented in **Table 1**. Acorns of three species (*Q. libani, Q. petraea* subsp. *iberica* and *Q. infectoria* subsp. *infectoria*) were germinated in plastic cups filled with water in room temperature [28]. The other 14 *Quercus* taxa studied (*Q. robur, Q. hartwissiana, Q. macranthera* subsp. *syspirensis, Q. frainetto, Q. vulcanica, Q. infectoria* subsp. *boissieri, Q. pubescens, Q. virgiliana,*

Species	Section	Locations
Q. coccifera	Ilex	Uşak, Hatay
Q. ilex	Ilex	Zonguldak, Düzce
Q. aucheri	Ilex	Aydın, İzmir, Muğla
Q. cerris	Cerris	Balıkesir
Q. ithaburensis	Cerris	Çanakkale, Balıkesir
Q. libani	Cerris	Between Erzincan-Tercan
Q. trojana	Cerris	Uşak University/1 Eylül Campus/Uşak
Q. petraea subsp. *iberica*	Quercus	Samsun: 2–3 km on the Ladik road after Havza
Q. infectoria subsp. *infectoria*	Quercus	Sakarya:Bilecik road to Taraklı from Geyve
Q. infectoria subsp. *boissieri*	Quercus	Between Dikili-Candarlı/İzmir
Q. pubescens	Quercus	Bayat/Afyon
Q. robur	Quercus	Uşak University/1 Eylul Campus/Uşak
Q. vulcanica	Quercus	Sultan Mountain/Afyonkarahisar
Q. hartwissiana	Quercus	Between Bursa-Yalova/Güney village
Q. frainetto	Quercus	Between Bursa-Yalova/Güney village
Q. macranthera subsp. *syspirensis*	Quercus	Between Abant-Mudurnu/Bolu
Q. virgiliana	Quercus	Abant lake/Bolu

Table 1. Species, sections and localities of studied species.

Q. ilex, Q. coccifera, Q. aucheri, Q. trojana, Q. cerris and *Q. ithaburensis* subsp. *macrolepis*) were germinated in the refrigerator at 4°C [29–31]. It can be stated that the method giving the best results for the germination of acorns was in the refrigerator at 4°C. As a first treatment, germinated roots in the lengths of 2–10 mm for each studied species were pretreated in α-monobromonaphthalene for about 16 h at 4°C. After first treatment, root tip meristems were fixed overnight with 3:1 absolute alcohol-glacial acetic acid mixture. Fixed root tips were stored in 70% alcohol at 4°C until analyses. Prior to staining, hydrolysis was carried out with 1 NHCl solution at 60°C for 13 min and 30 min depending on the species [28–31]. Root tips were then washed with distilled water. Finally, the root tips were stained with freshly prepared Feulgen or Orcein for 2 h. Squashes were made with 2% aceto orcein for Feulgen staining and 45% acetic acid for Orcein staining. The best metaphase plates were frozen in liquid nitrogen to make permanent using Entellan and then photographed 10 × 100. For all studied taxa, at least five plates of metaphase chromosomes were measured on the basis of long arm, short arm and arm ratio. Homologous chromosome pairs were identified and arranged. Chromosome pairs for all taxa were classified according to the nomenclature of Levan et al. [32] and Stebbins [33]. The karyotype asymmetry parameters like intrachromosomic asymmetric index (A_1) and interchromosomic asymmetric index (A_2) were calculated following Zarco [34].

3. Karyotype analyses of Turkish oaks

In the present chapter, karyotype analyses of the 16 oak species completed in previous studies from three sections in Turkey [28–31] and their taxonomic relationships with each other and European oaks were evaluated.

Section *Ilex* containing evergreen oaks in Turkey is represented by three species such as *Q. coccifera, Q. ilex* and endemic species *Q. aucheri* which are distributed only in Turkey and in some East Aegean islands of Greece. In this chapter, detailed chromosome measurements of all species from section *Ilex* are stated and compared with each other.

Section *Quercus* has the greatest number of species and widest distribution in the world. Similarly, in Turkey, most species belonging to the genus *Quercus* are in section *Quercus*: *Q. pontica, Q. robur, Q. hartwissiana, Q. macranthera* subsp. *sysprensis, Q. frainetto, Q. petraea, Q. vulcanica, Q. infectoria, Q. pubescens* and *Q. virgiliana* [2]. Chromosome analyses of all species from the section *Quercus* in Turkey have been completed except *Q. pontica*. In this chapter, all studied species are evaluated and compared with each other according to chromosomal parameters.

Section *Cerris* includes five species in Turkey: *Q. libani, Q. trojana, Q. cerris, Q. brantii* and *Q. ithaburensis* subsp. *macrolepis* [2]. All species examined except *Q. brantii* are evaluated in detail.

3.1. Comparisons of Turkish oaks on the basis of sections

3.1.1. Karyotype analyses and relations of oak species from section Ilex

Acorns as plant materials for each species were obtained from different locations in Turkey (**Table 1**). Analyzed somatic metaphase plates show that chromosomes of three taxa, namely *Q. coccifera, Q. ilex* and *Q. aucheri,* were very small and similar with diploid chromosome number 2n = 24 (**Table 2** and **Figure 1**).

Species	Somatic chromosome number	Karyotypic description	Length range (μm)	Haploid complement (μm)	A_1	A_2
Section: *Ilex*						
Q. coccifera	2n = 24	24m	(0.80–1.98)	14.61	0.19	0.27
Q. ilex	2n = 24	24m	(1.07–2.05)	17.47	0.21	0.20
Q. aucheri	2n = 24	24m	(1.12–2.56)	19.76	0.22	0.24
Section: *Cerris*						
Q. cerris	2n = 24	24m	(0.99–2.11)	17.33	0.18	0.23
Q. ithaburensis	2n = 24	24m	(0.90–2.06)	15.66	0.17	0.28
Q. libani	2n = 24	24m	(0.81–2.18)	16.53	0.19	0.29
Q. trojana	2n = 24	14m + 10sm	(2.29–6.65)	49.62	0.28	0.30
Section: *Quercus*						
Q. petraea subsp. *iberica*	2n = 24	24m	(0.86–1.66)	14.33	0.15	0.19
Q. infectoria subsp. *infectoria*	2n = 24	24m	(0.91–1.96)	16.17		0.22
Q. infectoria subsp. *boissieri*	2n = 24	24m	(1.02–2.35)	17.89	0.21	0.24
Q. pubescens	2n = 24	24m	(1.01–2.01)	16.89	0.19	0.21
Q. robur	2n = 24	24m	(1.75–3.92)	31.78	0.22	0.22
Q. vulcanica	2n = 24	24m	(1.25–3.13)	22.63	0.18	0.28
Q. hartwissiana	2n = 24	22m + 2sm	(0.85–1.83)	15.22	0.22	0.23
Q. frainetto	2n = 24	22m + 2sm	(0.76–1.80)	14.50	0.25	0.24
Q. macranthera subsp. *syspirensis*	2n = 24	22m + 2sm	(0.88–1.99)	16.04	0.22	0.21
Q. virgiliana	2n = 24	24m	(0.85–2.16)	15.84	0.22	0.27

Table 2. Species, somatic chromosome numbers, karyotypic descriptions and other morphometric parameters of previous analyses on the Turkish *Quercus* taxa.

Averages of chromosomal lengths of investigated species ranged from 0.80 to 2.56 μm. Among these, *Q. coccifera* has the smallest chromosome set (0.80–1.98) and haploid complement value with 14.61 μm. On the contrary, the highest chromosome set and haploid complement value were observed in *Q. aucheri* with 1.12–2.56 and 19.76, respectively. Among the studied taxa, the lowest intrachromosomal asymmetry index (A_1) and the highest interchromosomal asymmetry index (A_2) were observed in *Q. coccifera*. However, all investigated species showed very similar intrachromosomal asymmetry index (A_1).

Finally, it can be concluded that the members of *Ilex* section are similar in chromosomal parameters such as small chromosome set and haploid complement, 2n = 24 chromosomes, all metacentric chromosomes and very close A_1 values. The main reason of the similarity between

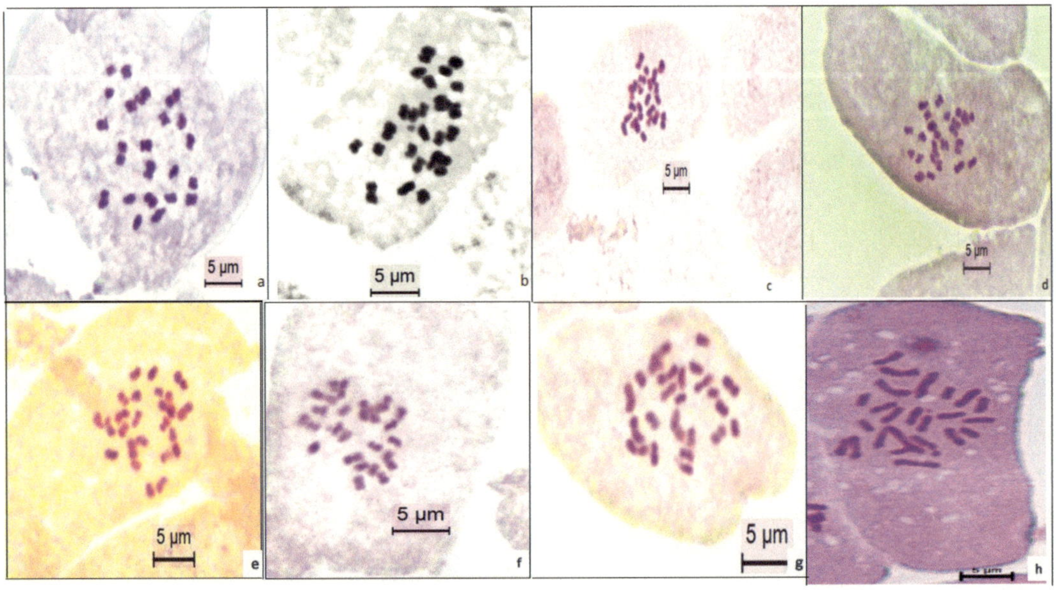

Figure 1. Somatic chromosomes of some Turkish oak species (a) *Q. hartwissiana,* (b) *Q. frainetto,* (c) *Q. macranthera* subsp. *syspirensis,* (d) *Q. virgiliana,* (e) *Q. infectoria* subsp. *boissieri,* (f) *Q. pubescens,* (g) *Q. vulcanica,* and (h) *Q. trojana* [30, 31].

species belonging to same section may be caused by the gene flow and genetic similarity. Besides that, in comparison between studied three taxa, it can be stated that *Q. ilex* and *Q. aucheri* show more similarity than *Q. coccifera* in all chromosomal parameters such as length range, haploid complement, A_1 and A_2 values. Similarly, Yılmaz et al. [35] stated in previous report on DNA comparison of three species belonging to *Ilex* section that *Q. ilex* and *Q. aucheri* were observed as close two separate groups and populations of *Q. coccifera* showed more differences than populations of *Q. ilex* and *Q. aucheri*.

Q. coccifera and *Q. ilex* analyzed before by D'emerico et al. [36, 37] show clearly differences with the presence of submetacentric chromosome pairs, while these two taxa examined in Turkey consist of all metacentric chromosomes and the less parametric values in all chromosomal morphometric measurements in comparison.

When we evaluate geographical distribution of these three taxa in Turkey.

Q. coccifera which is a Mediterranean element has the wide distribution area in comparison with *Q. ilex* and *Q. aucheri*. *Q. coccifera* is distributed along the coastal regions of the Mediterranean Sea, the Aegean Sea, the Marmara Sea and rarely the Black Sea.

Q. ilex which is another species of Mediterranean origin is limited to coastal regions of the Black Sea, the Marmara Sea and the Aegean Sea.

Q. aucheri which is an endemic species has not shown wide distribution and restricted to south-west Anatolia in Turkey.

Similarity observed between *Q. ilex* and *Q. aucheri* could be caused by distribution in more restricted area and more isolated habitats in comparison with *Q. coccifera*. Furthermore, hybridization behavior is mostly observed in oaks, especially in habitats where two or more species overlap [38, 39]. This situation may be reason of variation of *Q. coccifera*.

3.1.2. Karyotype analyses and relations of oak species from section Cerris

Plant materials for *Q. cerris, Q. ithaburensis* and *Q. trojana* were collected from West Anatolia in Turkey (**Table 1**). On the contrary of these three taxa, *Q. libani* was collected from East Anatolia region of Turkey (**Table 1**).

Detailed karyotype analyses of *Q. cerris, Q. ithaburensis, Q. libani* and *Q. trojana* from section Cerris were examined and determined the chromosome number for each taxon as 2n = 24 [28, 29, 31]. Chromosome length ranges of *Q. cerris, Q. ithaburensis* and *Q. libani* were very similar with (0.99–2.11), (0.90–2.06) and (0.81–2.18), respectively. Similarly, haploid complement values for these three taxa were quite close and compatible values with (17.33), (15.66) and (16.53), respectively. Karyotypic description of all taxa analyzed except *Q. trojana* consists of metacentric chromosomes. Furthermore, chromosomal asymmetry index, A_1 and A_2, were compatible for these three taxa like other chromosomal parameters. Finally, it can be stated that all taxa analyzed from section *Cerris* showed similar and low parametric values except *Q. trojana* (**Table 2**).

Q. trojana, contrary to other species, showed the highest values in all morphometric parameters such as length range, haploid complement, karyotypic description and A_1 and A_2 values. In other words, among the studied taxa, the biggest variation was determined in *Q. trojana*. Chromosome length range and haploid complement value of this taxon have the highest values with 2.29–6.65 and 49.62 μm, respectively, in comparison with other studied taxa. *Q. trojana* approximitaly show equal sum of metacentric and submetacentric chromosomes. Karyotypic description of this taxon consists of 14 metacentric and 10 submetacentric chromosomes (14m + 10sm) and shows similarity with the results provided from D'emerico et al. [36] (8m + 4mSC + 10sm + 2smSC). Chromosomal asymmetry index, A_1 and A_2, have the highest value among the examined species with 0.28 and 0.30, respectively.

The most important reasons that affect the species number, variation and distribution in different regions are their location, geomorphologic structure and climatic effects. Examined *Q. trojana* is a species belonging to the Uşak/Uşak University-Campus location. The location of *Q. trojana* is quite rich on account of species diversity. This taxon is located as mixed oak populations with *Q. cerris, Q. ithaburensis, Q. robur, Q. infectoria* and *Q. coccifera* in the same location. Hybridization is mostly observed in restricted zones where the habitats of two or more species overlap [38, 39]. Many oak species are located at the same region or even at the same location due to the factors mentioned earlier and this is one of the most important factors that increases the hybridization especially between species belonging to the same section. This situation may be a reason for the high variation in this taxa.

Especially north-west Turkey is one of the regions having the highest species diversity and distribution for oaks. Northwest Turkey contains oak species which range from 13 to 15 [4]. The main reason why this region is rich in oak variation is that it is a transitional zone between Asia and Europe. Anatolia has served as a migration route facilitating the penetration of Asiatic plant elements into Southeast Europe [12]. Turkey has been under the influence of numerous climatic regions and three phytogeographic regions (Euro-Siberian, Irano-Turanian and Mediterranean regions) due to its geomorphologic structure [4]. Another reason of the high species diversity for northwest Turkey is that it is the place where the two different phytogeographical regions (Euro-Siberian and Mediterranean regions) overlap. Locations of *Q. cerris* and *Q. ithaburensis* examined in this study and rich species diversity belong to the same region in northwest Turkey (**Table 1**). The main reason for the similarity between

Q. cerris and *Q. ithaburensis* belonging to same section may be because of the gene flow and genetic similarity. Furthermore, it can be stated that the reason for the similarity in their chromosomal parameters may be probably caused by hybridization between *Q. ithaburensis* and *Q. cerris* which is a very common oak in Turkey.

In comparison with European oaks, all parametric values provided from *Q. cerris* by D'emerico et al. [36] are higher than examined taxon in Turkey except chromosome number (2n = 24). These differences can be caused by different geographical regions having different oak populations and environmental conditions.

Q. libani is distributed along and in the eastern part of the Anatolian Diagonal. The eastern part of Anatolian Diagonal is known for its high mountains. Observations by Uslu and Bakış [4] have supported that the number of samplings is decreased in high mountain region of 1100–1200 m. Furthermore, human impact over the vegetation such as heavy grazing and forest destruction may be the other reason for the least diversity in this region. Distribution of species belonging to Section *Cerris* has a relation with Anatolian Diagonal. *Q. cerris, Q. ithaburensis* and *Q. trojana* have distribution in the western part of the Diagonal. On the contrary, *Q. libani* and the last species of the section, *Q. brantii,* have distribution in the eastern part of the Diagonal. Cytogenetic study made on *Q. brantii* can be very useful in understanding the relations between *Q. libani* and *Q. brantii.*

3.1.3. Karyotype analyses and relations of oak species from section Quercus

Section *Quercus* is characterized by 10 species in Turkey, and detailed karyotype analyses have been completed in all species such as *Q. robur, Q. hartwissiana, Q. macranthera* subsp. *syspirensis, Q. frainetto, Q. petraea, Q. vulcanica, Q. infectoria, Q. pubescens* and *Q. virgiliana* except *Q. pontica* [28, 30, 31]. *Q. infectoria* is represented by two subspecies known as *Q. infectoria* subsp. *infectoria* and *Q. infectoria* subsp. *boissieri* in this study. In other words, the section *Quercus* was represented by 10 taxa for cytogenetic comparison.

The chromosome number of 10 taxa analyzed from section *Quercus* support that the basic chromosome number of the genus is n = 12 (**Table 2** and **Figure 1**). Majority of the examined taxa in the section *Quercus* have karyotypes with predominance of metacentric chromosomes. However, it is observed that *Q. hartwissiana, Q. frainetto* and *Q. macranthera* subsp. *syspirensis* have two submetacentric chromosomes.

Plant samples of *Q. hartwissiana* and *Q. frainetto* were collected from the same location. Among the examined taxa, the karyotypes of *Q. hartwissiana* and *Q. frainetto* show close similarity in terms of chromosome number (2n = 24), chromosome morphology (22m + 2sm), small chromosome sets (0.85–1.83 and 0.76–1.80) and small haploid complement values with 15.22 and 14.50, respectively.

Q. hartwissiana, Q. frainetto, Q. macranthera subsp. *syspirensis* that are endemic taxa and *Q. virgiliana* were collected from the north-west region of Turkey having the highest species diversity and distribution due to its geomorphologic structure and climatic effects. The main reason for the similarity between these taxa may be because of the gene flow and genetic similarity due to the mixed oak population in restricted area and hybridization because of weak reproductive barrier between oak species especially belonging to the same section.

It was previously stated by D'emerico et al. [36] that chromosome number of *Q. frainetto* is 2n = 24 and karyotypic description of this taxon was 14m + 2mSC + 6sm + 2smSC. Although chromosome number of examined species showed the consistency with 2n = 24, other parameters such as chromosome morphology, haploid complement, A_1 and A_2 showed less values and differences according to D'emerico et al. [36].

Detailed chromosome measurements of *Q. virgiliana* were previously reported by D'emerico et al. [36]. It was stated by D'emerico et al. [36] that karyotypic description and A_1 value of *Q. virgiliana* were 10m + 4mSC + 8sm + 2smSC and 0.35, respectively. While the similarity in terms of chromosome number (2n = 24) in comparisons with karyotypes was observed, chromosome morphologies differentiated according to total lengths of chromosomes and karyotypic description. In this study, *Q. virgiliana* has the smaller chromosome set, all metacentric chromosomes and the less parametric value for A_1.

The less values for many measured parameters were observed in *Q. frainetto* and *Q. virgiliana* in comparison with D'emerico et al. [36]. These differences can be caused by oak species living in different geographical regions, hybridization and gene flow between oak species distributed in this area.

Q. pubescens has a wide distribution range in the northern, western, southern and central parts of Turkey. However, it is known as a species that has not crossed the eastern border of the Anatolian Diagonal. The *Q. pubescens* chromosome number was observed as 2n = 24 with all metacentric chromosomes. *Q. pubescens* has the small chromosome set, 1.01–2.01 μm, and haploid complement value, 16.89 μm. Chromosomal asymmetric index values, A_1 and A_2, show low value among the studied taxa with 0.19 and 0.21, respectively. Previously, chromosome numbers and morphometric parameters of *Q. pubescens* were reported by D'emerico et al. [37]. The chromosome number of this taxon was reported as 2n = 24 with 18 metacentric and 6 submetacentric chromosomes by D'emerico et al. [37]. In comparison with D'emerico et al. [37], chromosome number is compatible with 2n = 24 but chromosome morphologies showed differences with all metacentric chromosomes. However, as a result of a small chromosome set (1.01–2.01 μm), there were also differences in haploid complement value in comparison with the value reported (27.28). In Turkey, there are many hybrids of *Q. pubescens*, especially with *Q. petraea, Q. infectoria* and *Q. macranthera* subsp. *syspirensis* [2]. These differences can be caused by oaks living in different geographical regions and hybridization behaviors seen commonly between oak species living in mixed populations.

Quercus infectoria containing two subspecies such as *Q. infectoria* subsp. *infectoria* and *Q. infectoria* subsp. *boissieri* in Turkey has the widest distribution area, especially in the west, south and south-east regions of Turkey. *Q. infectoria* subsp. *infectoria* has a more limited distribution area compared to *Q. infectoria* subsp. *boissieri* in Turkey. In the comparison with these two subspecies, very similar results are observed with chromosome number and all metacentric chromosomes. Furthermore, it was observed that the chromosome sets of *Q. infectoria* subsp. *boissieri* and *Q. infectoria* subsp. *infectoria* were quite small and similar, 1.02–2.35 and 0.91–1.96, respectively.

Finally, it can be stated that all taxa anayzed belonging to section *Quercus* in Turkey are compatible with each other according to many chromosomal parameters such as length range, haploid complement, A_1 and A_2 value except *Q. vulcanica* and *Q. robur* (**Table 2**).

Q. robur together with *Q. vulcanica* show differences in chromosome lengths and haploid complement compared to the other examined taxa (**Table 2**). The haploid complement value of *Q. robur* is the highest with 31.78 and chromosome lengths range from 1.75 to 3.92 μm. The total chromosome number and karyotypic description for this taxon are compatible with other taxa. *Q. robur* is a species belonging to the Uşak/Uşak University-Campus location (**Table 1**). This taxon is located as mixed oak populations with *Q. cerris, Q. ithaburensis, Q. trojana* and *Q. coccifera* in the same location. Hybridization in restricted zones, where there is quite rich species diversity, was frequently observed. The reason for high variation may be because of its distribution with other taxa.

Q. vulcanica is an endemic taxon distributed in restricted areas such as Isparta/Eğirdir and Afyon/Sultan Mountains in Turkey. The haploid complement has the second highest value with 22.63 μm after *Q. robur* and chromosomal lengths ranged from 1.25 to 3.13 μm. Among the studied taxa, *Q. vulcanica* has the second lowest A_1 value (0.18) and the highest A_2 value (0.28) (**Table 2**). In other words, it can be stated that the endemic species *Q. vulcanica* exhibited high variation among the studied taxa in chromosome lengths, haploid complement value and asymmetric index A_1 and A_2.

This species is naturally distributed from 1200 to 2000 m altitude in restricted areas such as Kutahya-Turkmen Mountains, Konya-Sultan Mountains and Isparta-Eğirdir (Yukari Gokdere village). This high variation could be caused by the geographical distribution in this restricted area and more isolated habitats when compared with other oak species. *Quercus vulcanica* has been faced with the threat of extinction because of over exploitation for wooden home appliances, veneer and furniture. To protect this valuable resource, 1300.5 ha area near the Eğirdir Yukari Gökdere village was declared as a Nature Reserve Area for this endemic species. Eğirdir-Yukarıgökdere location is accepted as the best location of the species.

High variation observed in *Q. vulcanica* could be caused by distribution in restricted area and more isolated habitats in comparison with other oak species.

3.2. Situations that increase the variation and taxonomic problems among oak species in Turkey

- Hybridization and introgression are an important process in evolution, diversification and speciation of many plants [40, 41]. Oaks have long been considered a group with high frequency of widespread hybridization. Many morphological studies on *Quercus* species show the presence of hybridization with hybrid individuals that exhibit intermediate morphological features between parent taxa and support interspecific gene flow [14]. Furthermore, recently, many molecular techniques have been used to understand the relations between *Quercus* taxa and the determining of hybridization. Especially in Europe, species that dominate forests such as *Q. petraea, Q. robur* and *Q. pubescens* have been studied extensively using different molecular techniques [42–46]. However, oak taxonomy is still problematic and under debate often due to insufficient diagnostic morphological characters [24–25], weak reproductive barriers between species, wind-pollination [14, 22–25] and the lack of investigations for each taxon such as ecological, historical and genetic descriptors [25]. All these factors that is stated increase the taxonomic problems and make problematic the genus. In addition to these factors, identification of oak species can be difficult due to species boundaries being fuzzy as a result of interspecific gene flow and variation within species. The most important reason for hybridization in oaks is the presence of hybrid zones that are frequently observed when species separated by weak reproductive barriers come into geographical contact [47]. In other words, sympatry creates high opportunity for hybridization in oaks. Nevertheless, the climate selection for species that create the hybrid zones is very important because they determine the extent to which hybrids persist in a given locality. In Turkey, these factors that increase the hybridization and gene flow between oak species are frequently observed and cause taxonomic problems.

- Another situation affecting variation, hybridization, speciation and taxonomic problems in oaks is Anatolian Diagonal which separates Turkey into West and East (**Figure 2**). It can be stated that distributions of some *Quercus* species at the subspecies level is related to phytogeography and Anatolian Diagonal [4, 12, 48]. For example, while *Q. petraea* subsp. *pinnatiloba* has a distribution area in the eastern part of the Anatolian Diagonal, the other two subspecies of *Q. petraea, Q. petraea* subsp. *iberica* and *Q. petraea* subsp. *Petraea,* show distribution in the western part of the Diagonal. Similarly, *Q. robur* subsp. *robur* shows distribution in the western part of the Diagonal whereas *Q. robur* subsp. *pedunculiflora* has distribution in the eastern part of Diagonal. This separation at the subspecies level shows the importance of the Anatolian Diagonal in the evolution, diversification and speciation of many plants in Turkey. Furthermore, distributions of some oak species have considerable relation with Anatolian Diagonal in species level. Especially, distributions of species belonging to the Section *Cerris* show high relation with Anatolian Diagonal. While *Quercus brantii* and *Q. libani* have a distribution area in the eastern part of the Diagonal, *Q. ithaburensis* subsp. *macrolepis, Q. cerris* and *Q. trojana* show distribution in the west part of the Diagonal. However, some taxa from Section *Quercus* show distribution tendency with respect to the Diagonal. This is a case showing the influence on some oak species distribution of Anatolian Diagonal and as a result of this, gene flow is affected in intraspecific and interspecific level.

- Geomorphological structure and location of Turkey have an important effect on oak distribution and diversity. One of the main reasons why Turkey is rich in oak variation and distribution is that it is a transitional zone between Asia and Europe [4, 31]. Anatolia has served as a migration route facilitating the penetration of Asiatic plant elements into Southeast Europe [12]. Furthermore, Turkey has been under the influence of three different phytogeographic regions known as Euro-Siberian, Irano-Turanian and Mediterranean regions due to its geomorphologic structure [4, 31]. These phytogeographic regions that are caused by geomorphologic structure create different climatic regions that have an effect on species distribution, variation and the extent of hybrid zones.

- The changes that started at the beginning of the neotectonic period in the middle of the Miocene [49] affected the paleogeography of Turkey and have changed the geomorphology of Anatolia. Furthermore, the formation of today's topography together with climatic changes affecting the distribution of plant flora in Turkey has occured in the Quaternary, especially at the beginning of the Holocene. In Turkey, the main reason for oak richness, variation and distribution that is creating the hybrid zones is the geomorphology and climatic structure affected by geologic history.

- Another important reason that makes it difficult to understand the relationships among the oaks and increases the taxonomic problems in Turkey is the lack of adequate conservation programs for the use of oak trees. Oak trees have high economic value and have been used for many purposes, such as ornaments, wood, fuel wood, nonwood products, in the timber industry, and so on. Beside this, large parts of the forests in Turkey are being degraded by villagers due to excessive cutting and grazing. However, there is not enough protection strategies and management plans for oak species except one national park for *Quercus vulcanica*.
- The lack of comprehensive studies on the genus in Turkey is the other situation that is causing taxonomic problems and difficulties in understanding the relations between oak species.
- All studies on cytogenetic of Turkish oaks show that average chromosome lengths of oak species analyzed are below 2 μm. The effects on the chromosome lengths of chemicals used to obtain metaphase chromosomes during cytological studies considerably complicate cytological comparison in species with very small chromosomes such as oaks.

3.3. General comparisons between Turkish and European oaks on the basis of chromosome structures

Chromosome analyses of 16 *Quercus* species in Turkey have been completed [28–31] and confirmed that the chromosome number of the genus is consistent with 2n = 24 (**Table 2**). These results are in agreement with previous researches based on chromosome number of the *Quercus* species from different parts of the world [36, 37, 50–52]. However, some exceptions on different chromosome number are reported, contrary to results provided from studies on Turkish and European oaks [51, 53, 54]. Furthermore, occasionally ploidy variation may

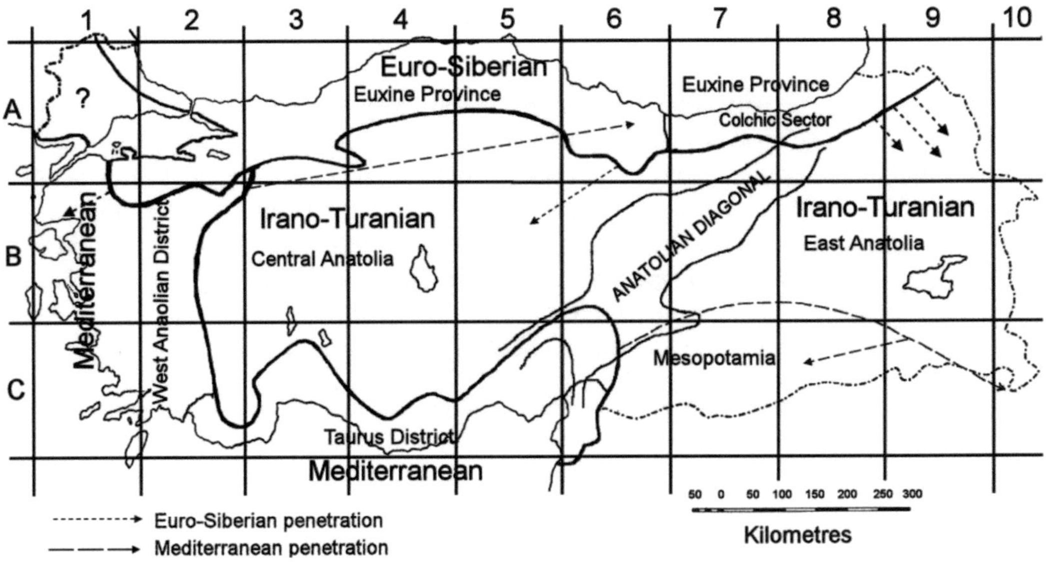

Figure 2. Anatolian Diagonal and phytogeographical regions of Turkey after Davis [4, 12].

be observed in individual trees such as triploid samples of *Q. robur* with 2n = 3x = 36 [55]. Similarly, the presence of endopolyploid cells (2n = 4x = 48) together with diploid cells is reported in an individual tree of *Q. frainetto* [36].

If the chromosome lengths in the evaluation of European and Turkey oaks are taken into consideration, it can be said that European oak species clearly have a bigger chromosome set than Turkish oak species. In addition to haploid complements and length ranges of species examined, it is observed that other morphometric parameters of karyotypes such as A_1 and A_2 show bigger values in opposition to Turkish oaks. Also, while the *Quercus* taxa analyzed in Turkey have predominant karyotypes with metacentric chromosomes, karyotypes having approximately equal sum of metacentric and submetacentric chromosomes are observed in European oaks. Consequently, higher values for many chromosomal parameters in European oaks than in Turkish oaks were observed. This situation may be caused by gene flow between different oak species because of weak reproductive barriers in different geographical regions.

Mixed populations that are composed of different oak species in different geographical regions and gene flow between these may be the cause of these differences between Turkish and European oaks. Also, differences of environmental and climatic factors affecting species variation and distribution in different geographical regions may be another important reason of this situation.

4. Conclusion

This study contributes to understanding the relations on the basis of sections among 16 Turkish oak species belonging to three sections and contributes to the comparisons between Turkish oaks and European oaks. Additionally, results obtained in this study provide useful knowledge on the cytogenetic of the genus *Quercus*. However, lack of adequate conservation programs for the use of oak trees in Turkey is the main reason that makes it difficult to understand the relationships among the oaks and increases the taxonomic problems. For example, endemic species *Q. vulcanica* has been faced with the threat of extinction because of overexploitation for wooden home appliances, veneer and furniture. To protect this valuable resource, 1300.5 ha near the Eğirdir Yukari Gokdere village was declared as a Nature Reserve Area for *Q. vulcanica*. Nevertheless, there is not enough protection for oak species except Nature Reserve Area for this endemic species. Furthermore, conservation of oak biodiversity should be considered not only in protected areas but also in managed forests as well.

Author details

Aykut Yılmaz

Address all correspondence to: aykutyilmaza@gmail.com

Faculty of Science and Arts, Department of Molecular Biology and Genetics, Uşak University, Uşak, Turkey

References

[1] Govaerts R, Frodin DG. World Checklist and Bibliography of Fagales (Betulaceae, Corylaceae, Fagaceae and Ticodenraceae). Great Britain: Royal Botanic Gardens, Kew; 1998

[2] Yaltirik F. Türkiye meşeleri teşhis kılavuzu. İstanbul: Yenilik Basımevi; 1984

[3] Hedge IC, Yaltırık F. *Quercus* L. In: Davis PH, editor. Flora of Turkey and the East Aegean Islands. Vol. 7. Edinburgh: Edinburgh University Press; 1982. pp. 659-683

[4] Uslu E, Bakış Y. Geographic distribution of Turkish Oaks. Dendrobiology. 2012;**67**:41-48

[5] Atalay İ. Kuvarterner'deki İklim Değişmelerinin Türkiye Doğal Ortamı Üzerindeki Etkileri. Türkiye Kuvaterner Sempozyumu, İTÜ Avrasya Yer Bilimleri Enstitüsü; 2-5 Haziran 2005; İstanbul. pp. 121-127

[6] Atalay İ. The Paleogeography of the Near East from Late Pleistocene to Early Holocene and Human Impact. İzmir: Ege University Press; 1992

[7] Atalay İ. Türkiye Vejetasyon Coğrafyası. Ege Coğrafya Dergisi. 1994;**2**:31-47

[8] Atalay İ. Effects of the climatic changes on the vegetation in the Near East. Bulletin Egyptian Geography Society. 1995;**68**:157-177

[9] Atalay İ. Relic and endemic plant communities and their importance in terms of climatic changes in Anatolia. In: Proceeds Int. Symp. İstanbul, Turkey: Earth System Sciences; 2004. p. 113

[10] Bottema S, Woldring H. Late quaternary vegetation and climate of southwestern Turkey. Paléo. 1984;**26**:123-149

[11] Zeist WV, Bottema S. Late quarternary vegetation of the near east. In: Beiheftezum Tübinger Atlas Des Vorderen Orients Reihe (Naturwissenschaften) Nr. Vol. 18. Wiebaden: Verlag; 1991

[12] Davis PH. Distribution patterns in Anatolia with particular reference to endemism. In: Davis PH, Harper PC, Hedge IC, editors. Plant Life of South West Asia. Edinburgh: Botanical Society of Edinburgh; 1971. pp. 15-27

[13] Çıplak B, Demirsoy A, Bozcuk AN. Distribution in Orthoptera in relation to the Anatolian Diagonal in Turkey. Art. 1993;**8**:1-20

[14] Borazan A, Babaç MT. Morphometric leaf variation in oaks (*Quercus*) of Bolu, Turkey. Annales Botanici Fennici. 2003;**40**:233-242

[15] Uslu E, Bakış Y, Babaç MT. A study on biogeographical distribution of Turkish oak species and their relations with the Anatolian diagonal. Acta Botanica Hungarica. 2011;**53**: 423-440

[16] Stace CA. Plant Taxonomy and Biosystematics. Cambridge: Cambridge University Press; 1989

[17] Jensen RJ. Assessing patterns of morphological variation of *Quercus* spp. in mixed oak communities. American Midland Naturalist. 1988;**120**:120-135

[18] Bakış Y, Babaç MT. Morphological variability of acorns and its taxonomic significance in *Quercus* L. from Turkey. Bangladesh Journal of Botany. 2014;**43**(3):293-299

[19] Yılmaz A, Uslu E, Babaç MT. Morphological variability of evergreen oaks (*Quercus*) in Turkey. Bangladesh Journal of Plant Taxonomy. 2017;**24**(1):39-47

[20] Kremer A, Petit RJ. Gene diversity in natural populations of oak species. Annals of Forest Science. 1993;**50**:186-202

[21] Hokanson SC, Isebrands JG, Jensen RJ, Hancock JF. Isozyme variation in oaks of the Apostle Islands in Wisconsin: Genetic structure and levels of inbreeding in *Quercus rubra* and *Quercus ellipsoidalis* (Fagaceae). American Journal of Botany. 1993;**80**:1349-1357

[22] Bacilieri R, Ducousso A, Petit RJ, Kremer A. Mating system and asymmetric hybridization in a mixed stand of european oaks. Evolution. 1996;**50**:900-908

[23] Manos PS, Doyle JJ, Nixon KC. Phylogeny, biogeography, and processes of molecular differentiation in *Quercus* subgenus *Quercus* (Fagaceae). Molecular Phylogenetics and Evolution. 1999;**12**:333-349

[24] Denk T, Grimm GW. The oaks of western Eurasia: Traditional classifications and evidence from two nuclear markers. Taxon. 2010;**59**(2):351-366

[25] Simeone MC, Piredda R, Papini A, Vessella F, Schirone B. Application of plastid and nuclear markers to DNA barcoding of Euro-Mediterranean oaks (*Quercus*, Fagaceae): Problems, prospects and phylogenetic implications. Botanical Journal of the Linnean Society. 2013;**172**:478-499

[26] Petit RJ, Bodenes C, Ducousso A, Roussel G, Kremer A. Hybridization as a mechanism of invasion in oaks. New Phytologist. 2003a;**161**:151-164

[27] Oh SH, Manos PS. Molecular phylogenetics and cupule evolution in Fagaceae as inferred from nuclear CRABS CLAW sequences. Taxon. 2008;**57**:434-451

[28] Yılmaz A, Uslu E, Babaç MT. Karyological studies on four *Quercus* L. species in Turkey. Caryologia. 2008;**61**(4):397-401

[29] Yılmaz A, Uslu E, Babaç MT. Cytogenetic studies on *Quercus* L. (Fagaceae) species belonging to *Ilex* and *Cerris* sections in Turkey. Caryologia. 2011;**64**(3):297-301

[30] Yılmaz A. Cytotaxonomic study of *Quercus* L. species from section *Quercus* in Turkey. Caryologia. 2017;**70**(2):141-146

[31] Yılmaz A. Karyomorphology of some *Quercus* L. species from section *Quercus* and *Cerris* in Turkey. Caryologia. 2018; in press

[32] Levan A, Fredga K, Sandbdberg AA. Nomenclature for centromeric position on chromosomes. Hereditas. 1964;**52**:201-220

[33] Stebbins GL. Chromosomal Evolution in Higher Plants. London: Edward Arnold; 1971

[34] Zarco CR. A new method for estimating karyotype asymmetry. Taxon. 1986;**35**(3):526-530

[35] Yılmaz A, Uslu E, Babaç MT. Molecular diversity among Turkish oaks (*Quercus*) using random amplified polymorphic DNA (RAPD) analysis. African Journal of Biotechnology. 2013;**12**:6358-6365

[36] D'emerico S, Bıanco P, Medaglı P, Schırone B. Karyotype analysis in *Quercus* ssp. (Fagaceae). Silvae Genetica. 1995;**44**:2-3

[37] D'emerico S, Pacıolla C, Tomması F. Contribution to the karyomorphology of some species of the genus *Quercus*. Silvae Genetica. 2000;**49**:6

[38] Muller CH. Ecological control of hybridization in *Quercus*: A factor in the mechanism of evolution. Evolution. 1952;**6**:147-161

[39] Rushton BS. Natural hybridization within the genus *Quercus* L. Annales des Sciences Forestières. 1993;**50**:18

[40] Arnold ML. Natural Hybridization and Evolution. New York. USA: Oxford University Press; 1997

[41] Rieseberg LH, Willis JH. Plant speciation. Science. 2007;**317**:910-914

[42] Dumolin-Lapegue S, Kremer A, Petit RJ. Are chloroplast and mitochondrial DNA variation species-independent in oaks? Evolution. 1999;**53**:1406-1413

[43] Bruschi P, Vendramin GG, Busotti F, Grossoni P. Morphological and molecular differentiation between *Quercus petraea* (Matt) Liebl. and *Quercus pubescens* Willd. (Fagaceae) in northern and central Italy. Annals of Botany. 2000;**85**:325-333

[44] Bordacs S, Popescu F, Slade D, Csaıkl UM, Lesur I, Borovıcs A, Kezdy P, Könıg AO, Gümöry D, Brewer S, Burg K, Petıt RJ. Chloroplast DNA variation of oaks in northern Balkans and in the Carpathian Basin. Forest Ecology and Management. 2002;**156**:197-209

[45] Petit RJ, Csaikl U, Bordács S, Burg K, Coart E, Cottrell J, van Dam B, Deans JD, Glaz I, Dumolin-Lapègue S, Fineschi S, Finkeldey R, Gillies A, Goicoechea PG, Jensen JS, König A, Lowe AJ, Madsen SF, Mátyás G, Munro RC, Olalde M, Pemonge M-H, Popescu F, Slade D, Tabbener H, Taurchini D, Ziegenhagen B, Kremer A. Chloroplast DNA variation in European white oaks: Phylogeography and patterns of diversity based on data from over 2600 populations. Forest Ecology and Management. 2002b;**156**:5-26

[46] Neophytou C, Douvani A, Aravanopoulos F. Conservation of nuclear SSR loci reveals hight affinity of *Quercus infectoria* ssp. *veneris* A. Kern (Fagaceae) to section Robur. Plant Molecular Biology Reporter. 2008;**26**:133-141

[47] Barton NH, Hewitt GM. Analysis of hybrid zones. Annual Review of Ecology and Systematics. 1985;**16**:113-148

[48] Ekim T, Güner A. The Anatolian Diagonal: Fact or fiction? Proceedings of the Royal Society of Edinburgh. 1986;**89B**:69-77

[49] Şengör AMC, White GW, Dewey JF. Tectonic evolution of the Bitlis suture, southeastern Turkey: Implications for the tectonics of Eastern Mediterranean. Rapport Commission international Mer Méditerranée. 1979;**25/26**(2a):95-97

[50] Ohrı D, Ahuja MR. Giemsa C-banded karyotype in *Quercus* L. (oak). Silvae Genetica. 1990; **39**:5-6

[51] Zoldos V, Pape D, Brown SC, Panaud O, Iljak-Yakovlev S. Genome size and base composition of seven *Quercus* species: Inter- and intra- population variation. Genome. 1998;**41**: 162-168

[52] Kurokawa Y, Yonezawa Y. Karyotype analysis of fifteen species of *Quercus* L. (Fagaceae) in Japan. Chromosome Science. 2004;**8**(4):209

[53] Chokchaıchamnankıt P, Anamthawat-Jonsson K, Chulalaksananukul W. Chromosomal mapping of 18S-25S and 5S ribosomal genes on 15 species of Fagaceae from northern Thailand. Silvae Genetica. 2008;**57**(1):5-13

[54] Ribeiro T, Loureiro J, Santos C, Morais-Cecilio L. Evolution of rDNA FISH patterns in the Fagaceae. Tree Genetics & Genomes. 2011;**7**(6):1113-1122

[55] Butorina AK. Cytogenetic study of diploid and spontaneous triploid oaks, *Quercus robur* L. Annales des Sciences Forestieres. 1993;**50**:144-150

Chapter 3

Application of Karyotype and Genetic Characterization Analyses for Hybrid Breeding of *Epinephelus* Groupers

Mei-Chen Tseng and Kuan-Wei Shih

Additional information is available at the end of the chapter

http://dx.doi.org/10.5772/intechopen.80414

Abstract

In this study, karyotypes and Cyt *b* gene sequences of seven different species of grouper including *Plectropomus leopardus*, *Epinephelus coioides*, *E. flavocaeruleus*, *E. fuscoguttatus*, *E. lanceolatus*, *E. polyphekadion*, and *E. tukula* were examined. All chromosome numbers from seven groupers were 2n = 48 with a high number of telocentric chromosomes (38–48) and fundamental arm numbers (FNs) (48–54). The mitochondrial Cyt *b* gene was used to establish the barcodes of seven groupers and analyze phylogenetic relationships among these species. We discovered that *Epinephelus* groupers should be classified as monophyly. The minimum genetic distance expressed between *E. coioides* and *E. tukula* was 0.1276. From results of the cytogenetic and molecular analyses, it was demonstrated that *Plectropomus* is a relatively primitive genus of grouper, while *Epinephelus* is a more-modern derived genus. Results also showed that *E. coioides* and *E. tukula* have similar genetic characters and karyotypes, and should be foremost considered for artificial hybridization strategies. Furthermore, information on karyotypes of species within the *Epinephelus* is still insufficient, and further elucidation of karyotypes of *Epinephelus* will be a great help to future genetic breeding research.

Keywords: barcode, cytochrome *b*, cytogenetic, genetic distance, hybridization

1. Introduction

Epinephelus groupers (Perciformes, Serranidae) are widely distributed in tropical and subtropical waters [1] and comprise 89 species (valid names) in marine habitats worldwide [2]. Most known grouper species are in the Indian-Pacific Ocean, 11 species along the West Atlantic coast, nine species in the East Atlantic Ocean and Mediterranean, and eight species in the eastern Pacific Ocean. Only a few groupers are distributed across different oceans [1]. Forty-one species of groupers in total were found in coastal waters of Taiwan [3].

Grouper is an important aquatic product in the world. In addition to abundant grouper caught at sea, the artificial breeding grouper is also a major aquatic product in the fishery trade. In the past, most grouper fry were from Southeast Asian countries such as the Philippines, Indonesia, and Thailand. However, survival rates markedly decreased due to catching and transportation. Nowadays, breeding techniques have been completely established for major commercial groupers, and so most grouper fry are bought from artificial breeding farms. Currently, *Epinephelus akaara, E. areolatus, E. awoara, E. bleekeri, E. bruneus, E. fuscoguttatus, E. lanceolatus, E. septemfasciatus, E. tauvina, E. coioides,* and *E. malabaricus* can be artificially reared and bred, especially *E. malabaricus*, which is the most successful case. Groupers have similar external morphologies, and their body color characteristics are not stable. Juveniles and adult fishes may show completely different color patterns. Therefore, it is often impossible to effectively distinguish species with similar morphologies in the adult stage [1, 4, 5]. As to their mating systems, incorrect identification of parents and progeny in rearing and breeding farms may cause artificial full-breeding plans and hybridization strategies to fail; moreover, this will result in significant fishery losses [1, 5, 6].

Traditionally, grouper species were classified using morphological and skeletal features [1, 7–9]. In the past two decades, molecular genetic technology has been dramatically developed and is now widely used in taxonomic and systematics studies. As Ref. [5] analyzed 42 species of grouper including three genera (*Epinephelus, Cephalopholis,* and *Mycteroperca*) using partial 16S ribosomal (r)DNA sequences. Results of that phylogenetic study revealed that both genera *Epinephelus* and *Mycteroperca* belong to the same clade, and it was inferred that Serranidae comprised a paraphyletic group.

Nowadays ichthyologists also use variable staining methods to obtain cytogenetic information of fish [10, 11]. According to previous studies, the number of chromosomes in groupers are 2n = 48, most of which are telocentric chromosomes, and fundamental numbers range 48–62 [12]. Some reports on the cytogenetics of grouper indicated that silver-binding nucleolar organizing regions (Ag-NORs) are highly conserved on the chromosome 24, but variations occur in the location between different groupers [13–18]. It is generally believed that such variations may be caused by an inversion of the arms during chromosome evolution. To study an evolutionary model of chromosomes and identify species, staining techniques were used often to analyze the karyotype and cytogenetics of groupers.

More than ten groupers have been successfully cultivated in Taiwan. However, most groupers have similar external morphologies, and their color patterns are quite unstable. Often grouper in different life stages exhibit inconsistent color distributions that resulted in the species identification of grouper fry being controversial or confusing [1, 5]. In the aquaculture industry, misidentification frequently occurs in different growth stages of groupers, and this can cause serious problems, such as chaos of market prices, interspecific ecological competition, and breeding strategy failures.

It is important to understand the karyotype and phylogeny of cultured grouper for a successful strategy of genetic breeding. That is when studying hybridization strategies of groupers, selecting similar karyotypes and closely related species for the parents may result in relatively higher success potential for hybridization. Therefore, the establishment of grouper karyotype and barcode data in this study will provide more-perfect genetic bases for species identification to improve possibilities for genetic breeding. The present study analyzed the mitochondrial cytochrome (Cyt) *b* gene sequences and chromosomal characters of seven cultured groupers in Taiwan. These results will provide farmers with more genetic information of groupers to develop useful breeding strategies for hybridization in the future.

2. Materials and methods

2.1. Sampling

Seven groupers, *Epinephalus lanceolatus, E. tukula, E. flavocaeruleus, E. polyphekadion, E. fuscoguttatus, E. coioides,* and *Plectropomus leopardus,* were collected from fish markets in Tungkang, southern Taiwan (**Figure 1**) for chromosome preparation and DNA sequence analysis. A piece of muscle tissue from each specimen was preserved in 95% ethanol (EtOH) and stored at the Fish Biology Lab in National Pingtung University of Science and Technology. Seven species were used for the karyotype analysis and Cyt *b* gene sequencing.

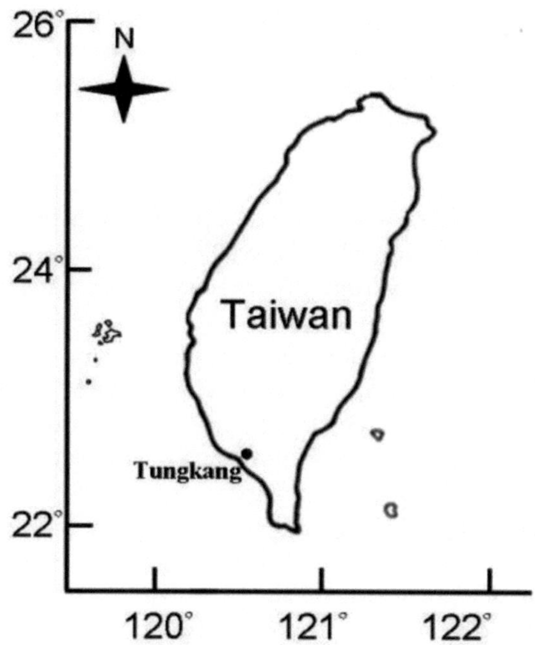

Figure 1. Sampling location of groupers.

2.2. Chromosomal preparation and karyotype analyses of groupers

The cell culture solution contained Eagle's minimal essential medium (MEM) with 15% fetal bovine serum and 0.0001% colchicine, followed by filter-sterilization (0.45 μm). Kidney tissue was cut and placed in the cell culture solution. The solution tubes were placed on a rotary shaker (100 rpm) and then incubated at room temperature for 2 h to allow cells to remain in the metaphase of the cell cycle. The cell culture solution was centrifuged at 3000 rpm for 5 min, and the supernatant was discarded. KCl (at 0.075 M) was added and allowed to sit at room temperature for 30 min. After centrifugation at 3000 rpm for 5 min, the supernatant was discarded, and a freshly prepared fixative solution (methanol: acetic acid = 3:1) was added at room temperature for 15 min. The mixture was centrifuged at 3000 rpm for 5 min, the supernatant was discarded, and this step was repeated two or three times. The cell suspension was dropped onto a heated glass slide and air-dried. After the slide had been stained with 5% Giemsa dye for 10 min, it was rinsed with water and air-dried. The slide was mounted and observed by microscopy.

In addition, some fresh chromosome slides were stained with $AgNO_3$. Two drops of 2% (w/v) gelatin and four drops of a 50% $AgNO_3$ solution were mixed and then dropped onto a slide with a cover glass. These slides were incubated at 70°C until they presented a yellowish-brown color. The slides were gently rinsed with double-distilled (dd)H_2O. After being air-dried at room temperature, the slides were mounted with gum arabic [19]. Chromosomes were observed with an optical microscope (Leica Microsystems, Wetzlar, Germany) (at 1000× with an oil lens). Digital images of the chromosomes were recorded and analyzed with a chromosome band analytical system (BandView 5.5, Applied Spectral Imaging, Migdal HaEmek, Israel). Chromosomes stained with Giemsa were classified into four groups, metacentric (m), submetacentric (sm), subtelocentric (st), and telocentric (t), according to the system described by [20]. Locations of chromosomes determined by $AgNO_3$ staining were observed and marked on photos.

2.3. DNA isolation

Approximately 100 mg of muscle tissue from each specimen was put into an Eppendorf tube. Before DNA purification, the tube was placed in a 60°C oven for 10 min to evaporate the EtOH. Genomic DNA was isolated using a Gentra Puregene Core kit A (Qiagen, Venlo, the Netherlands), and the purified DNA specimen was dissolved in TE buffer (1 M Tris–HCl at pH 8.0 and 0.2 mL EDTA, 0.5 M). DNA concentrations were estimated using a Nanodrop 2000C spectrophotometer (Thermo Fisher Scientific, Waltham, MA, USA) at an absorbance of 260 nm. The purity of DNA preparations was checked by the ratio of absorbances at 260 and 280 nm ($A_{260}/A_{280} \geq 1.8$). DNA stock solutions were stored in a − 20°C freezer.

2.4. Cyt *b* gene sequencing and analysis

In total, 50 μL of reactant of a polymerase chain reaction (PCR) contained 5 ng genomic DNA, 10 pmol each of the forward and reverse primers, 4 μL 2.5 mM dNTP, 0.2 μL 25 mM $MgCl_2$, 1 U *Taq* polymerase, and 5 μL 10× buffer, with ddH_2O added to 50 μL. The forward and reverse primers of the Cyt *b* gene were FOR (5′-CGAACGTTGATATGAAAAACCATCGTTG-3′) and UnvH (5′-ATCTTCGGTTTACAAGAC CGGTG-3′), respectively [6]. The Cyt *b* gene was amplified using a PCR machine (BIO-RAD MJ Mini Gradient Thermal Cycler, Conmall Biotechnology, Singapore) with initial denaturation at 95°C for 3 min; 35 cycles of 95°C for 1 min, 50°C for 1 min, and 72°C for 1 min; with a final extension of 72°C for 10 min. The reaction was cooled down to 25°C for 10 min. PCR products of the Cyt *b* gene were checked using 1% agarose gel electrophoresis and then stained with ethidium bromide (EtBr; 0.5 mg/mL). Target DNA fragments were eluted with a DNA Clean/Extraction kit (GeneMark, Taichung, Taiwan). Sizes of the purified DNA fragments were checked and then stored in a −20°C freezer. DNA fragments were directly sequenced on an Applied Biosystems (ABI, Foster City, CA, USA) automated ABI3730x1 DNA sequencer using a Bigdye sequencing kit (Perkin-Elmer, Wellesley, MA, USA). FOR or UnvH primers were used in the sequencing reaction, and the PCR cycle parameters for sequencing were 35 cycles of 30 s at 95°C, 30 s at 50°C, and 1 min at 72°C.

In total, seven Cyt *b* sequences were obtained in this study. Homologous sequences were aligned using ClustalW [21] and then manually checked. Interspecific genetic distances were analyzed using the Kimura-2-parameter (K2P) model [22], and numbers of different nucleotides were calculated with MEGA software [23]. The best-fitting models of DNA substitution were determined using the lowest Bayesian Information Criterion (BIC) scores [24]. The phylogenetic trees of Cyt *b* sequences were constructed using the Neighbor-joining (NJ) [25] and Maximum-likelihood (ML) methods [26]. Cluster confidence levels of Cyt *b* were assessed using a bootstrap analysis with 1000 replications [27].

3. Results

3.1. Karyotype analyses

In cytogenetic studies, Giemsa staining of seven groupers indicated that the diploid number of these species was 2n = 48. The karyotypic formulae were 2 sm + 46 t for *E. coioides, E. fuscoguttatus,* and *E. tukula*; 6 sm + 4 st + 38 t for *E. lanceolatus;* 2 st + 46 t for *E. flavocaeruleus;* 6 sm + 42 t for *E. polyphekadion;* and 48 t for *P. leopardus*. All of those specimens had a high number of telocentric chromosomes (38–48) and fundamental arm numbers (FNs) that ranged 48–54 (**Figure 2**, **Table 1**).

In Ag-NO_3 staining, four *Epinephelus* species (*E. coioides, E. fuscoguttatus, E. tukula,* and *E. lanceolatus*) and *P. leopardus* were completed. *Epinephelus coioides, E. fuscoguttatus,* and *E. tukula* had one pair of Ag-NORs located on the short arm of the sm chromosome; *E. lanceolatus* had two pairs of Ag-NORs located on the short arm of the sm chromosome; and *P. leopardus* had one pair of Ag-NORs, located near the centromere of larger telocentric chromosomes (**Figure 3**).

3.2. Cyt *b* sequence analysis

All Cyt *b* gene sequences from seven groupers were 1141 bp for *E. lanceolatus, E. tukula, E. flavocaeruleus, E. polyphekadion, E. fuscoguttatus, E. coioides,* and *P. leopardus*. Percentages of nucleotide compositions did not significantly differ among these *Epinephelus* species, as the A + T ratios were in the range of 52.1% (*E. flavocaeruleus*) - 56.7% (*E. polyphekadion*). Interspecific

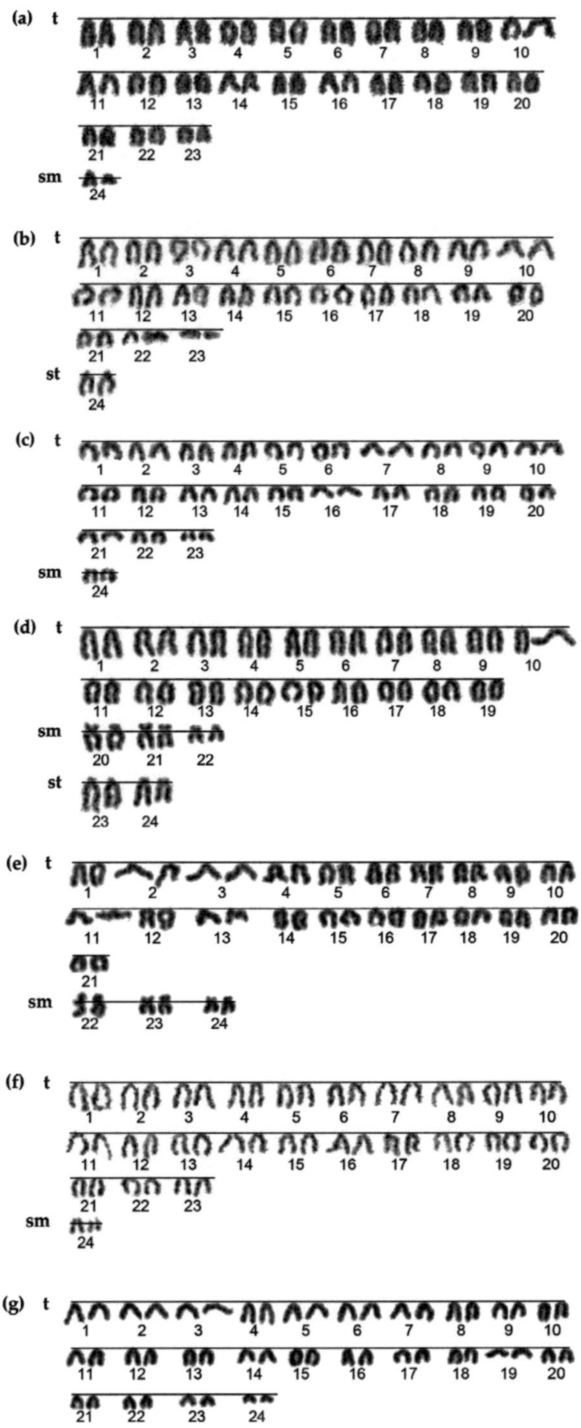

Figure 2. Karyotype analyses of seven groupers: (a) *Epinephelus coioides;* (b) *E. flavocaeruleus;* (c) *E. fuscoguttatus;* (d) *E. lanceolatus;* (e) *E. polyphekadion;* (f) *E. tukula;* and (g) *Plectropomus leopardus.*

Species	2*n*	FN	Formulae	Reference
E. adscensionis	48	48	48 a	[28]
E. akaara	48	48	5 st + 43 a	[40]
E. alexandrinus	48	48	48 a	[15]
E. awoara	48	48	48 a	[13]
E. bruneus	48	54	2 m + 4 sm + 42 a	[46]
E. caninus	48	48	48 a	[16]
E. coioides	48	50	2 sm + 46 a 2 sm + 46 t	[12] Present study
E. diacanthus	48	50	2 sm + 46 a	[41]
E. fario	48	62	4 m + 6 sm + 4 st + 34 a	[42]
E. fasciatomaculosus	48	48	48 a	[43]
E. fasciatus	48	48	48 a	[43]
E. faveatus	48	50	2 m + 46 a	[48]
E. flavocaeruleus	48	48	2 st + 46 t	Present study
E. fuscoguttatus	48	50	2 sm + 46 t	Present study
E. guaza	48	48	48 a	[15]
E. guttatus	48	48	48 a	[15]
E. lanceolatus	48	54	6 sm + 4 st + 38 t	Present study
E. malabaricus	48	48	48 a	[44]
E. marginatus	48	48	48 a	[18]
E. merra	48	62	4 m + 6 sm + 4 st + 34 a	[42]
E. moara	48	48	48 a 4 sm + 44 a	[45] [46]
E. polyphekadion	48	54	6 sm + 42 t	Present study
E. sexfasciatus	48	50	2 sm + 46 a	[47]
E. tauvina	48	50	2 sm + 46 a	[16]
E. tukula	48	50	2 sm + 46 t	Present study
P. leopardus	48	48	48 t	Present study

E., *Epinephelus*; *P.*, *Plectropomus*; 2*n*, diploid number; FN, fundamental number; metacentric (m), submetacentric (sm), subtelocentric (st), and telocentric (t), according to the system described in Ref. [20].

Table 1. Karyotype data of the Epinephelinae.

p-distances and K2P genetic distances ranged 0.1149 and 0.1284 (*E. tukula vs. E. coioides*) to 0.1814 and 0.2138 (*E. flavocaeruleus vs. E. polyphekadion*) (**Table 2**). The best model of nucleotide evolution was estimated to be the TN93 + G + I model with BIC = 9065.099. The NJ and ML analyses showed that *E. tukula* and *E. coioides* had a close phylogenetic relationship with extremely high bootstrap support (**Figure 4**). This result agreed with the hypothesis that *Epinephelus* is a monophyletic group.

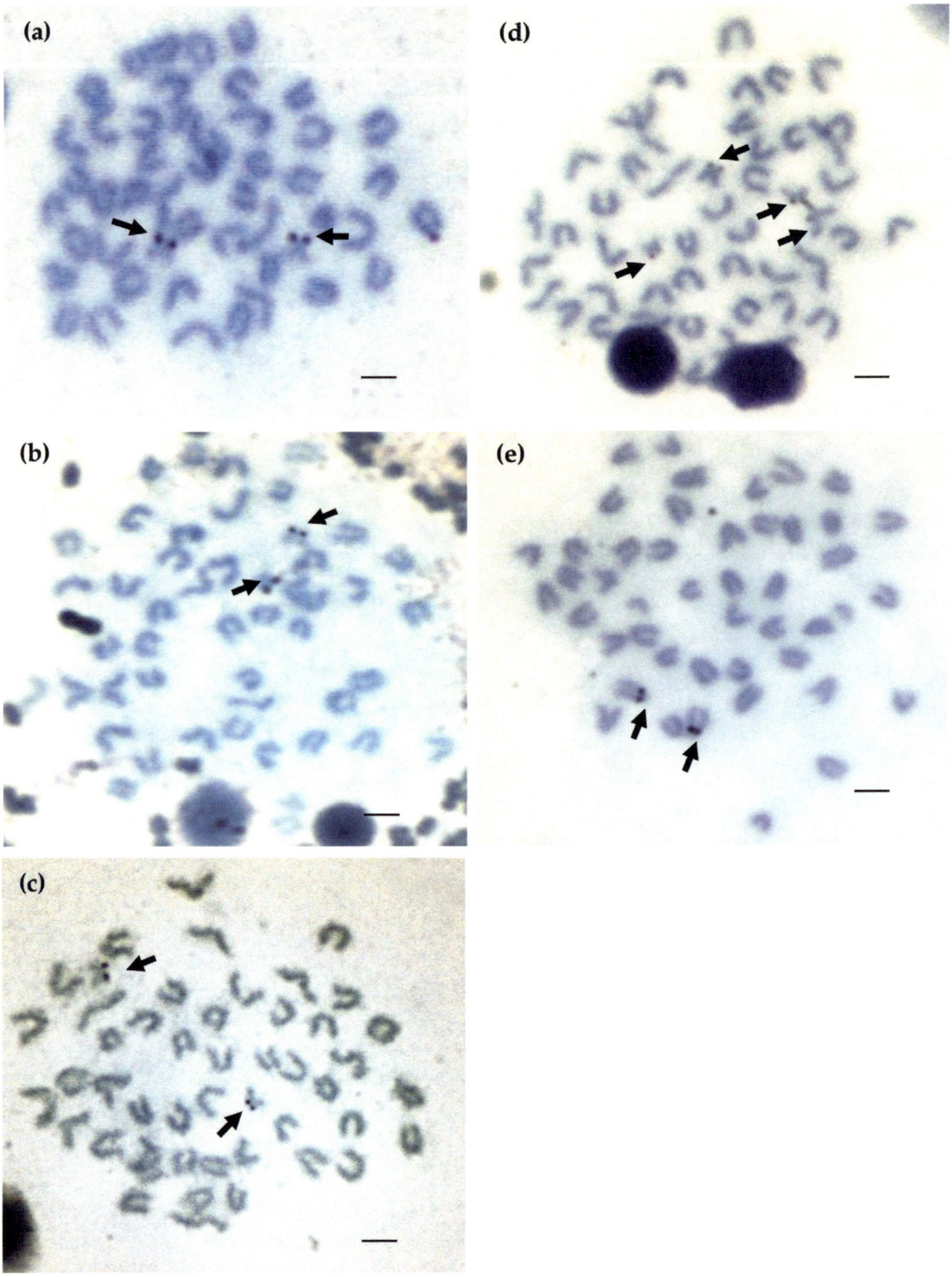

Figure 3. Silver-binding nucleolar organizing regions (Ag-NORs) results from five groupers: (a) *Epinephelus coioides*; (b) *E. fuscoguttatus*; (c) *E. tukula*; (d) *E. lanceolatus*; and (e) *Plectropomus leopardus*. The arrows indicate Ag-NORs. The bar equals 5 μm.

Code	Species name	1	2	3	4	5	6	7
1	*E. lanceolatus*	—	0.1422	0.1649	0.1474	0.1430	0.1333	0.2344
2	*E. tukula*	0.1635	—	0.1658	0.1684	0.1360	0.1149	0.2186
3	*E. flavocaeruleus*	0.1908	0.1934	—	0.1814	0.1578	0.1604	0.2272
4	*E. polyphekadion*	0.1690	0.1994	0.2138	—	0.1516	0.1595	0.2237
5	*E. fuscoguttatus*	0.1638	0.1561	0.1817	0.1751	—	0.1350	0.2123
6	*E. coioides*	0.1510	0.1284	0.1853	0.1858	0.1436	—	0.2307
7	*P. leopardus*	0.2859	0.2617	0.2738	0.2699	0.2529	0.2803	—

Table 2. *p*-distance genetic distances (above the diagonal) and Kimura 2-parameter distances (below the diagonal) of cytochrome *b* gene sequences among *Epinephelus* groupers and the outgroup *Plectropomus leopardus*.

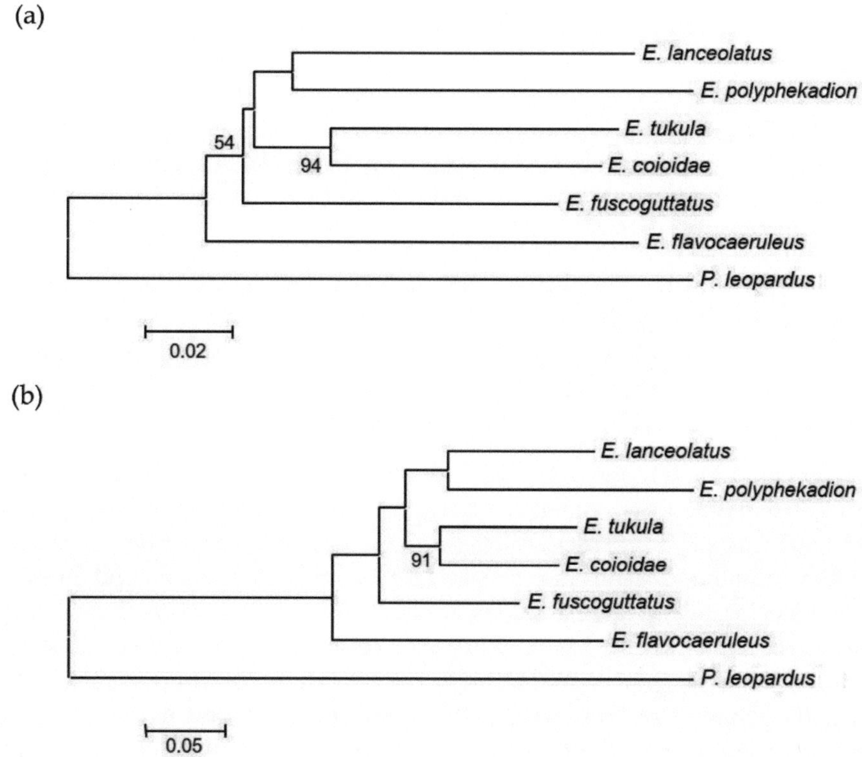

Figure 4. (a) The Neighbor-joining and (b) the Maximum-likelihood trees among *Epinephelus* species based on the cytochrome *b* gene analysis.

4. Discussion

In this study, Epinephelinae fish (*E. lanceolatus, E. tukula, E. flavocaeruleus, E. polyphekadion, E. fuscoguttatus, E. coioides,* and *P. leopardus*) showed a common synapomorphic character of chromosomal number, 2n = 48, and high numbers of telocentric chromosomes (38–48). By sorting out the cytogenetic information of 23 Epinephelinae species, it was found that chromosomal numbers of these groupers were 48, showing highly conserved characteristics, and FNs ranged 48–62, with more than half of these groupers exhibiting FN = 48 characteristics (**Table 1**), in accordance with conservative chromosomal morphological features described in Ref. [28]. In the other hand, variations in FNs are mainly caused by chromosomal rearrangements and play important roles in the speciation process [29].

In cytogenetic studies, karyotypes, FNs, Ag-NORs, and C-bands were demonstrated to have interspecific specificities, and many studies used these techniques to explore interspecific evolutionary relationships [30–32]. Currently, reports related to chromosomes of *Epinephelus* groupers worldwide are only available for 23 of 89 groupers; *e.g.*, karyotypes of *E. marginatus* were analyzed from three different sampling sites in the Mediterranean. Results showed chromosomal numbers of 2n = 48; conserved C-bands and Ag-NOR positions were observed on the 24th pair of chromosomes of specimens from all three samples, but those were also found on 2nd pair chromosomes of one specimen [18]. In order to confirm the above results, fluorescence in situ hybridization (FISH) was performed using 18S rDNA as a probe. Fluorescence reacted to the 2nd and 24th pairs of chromosomes confirming that a difference existed between samples. The authors reasoned that this may have been a species-specific manifestation, and further studies are required to confirm whether they can be population-specific markers.

Molecular phylogenetic analyses showed that both *Plectropomus* and *Cephalopholis* are more primitive genera than *Epinephelus* [5, 6, 12, 33, 34]. In this study, the chromosomal number of *P. leopardus* was 2n = 48 t. All current cytogenetic studies of *Epinephelus* groupers have shown that few of them are not composed of 2n = 48 t. These results support 48 t being an ancestral character of Serranidae fish [12], and *Epinephelus* groupers may be a later-derived genus.

In Ref. [12] observed three types of Ag-NORs distribution pattern: type I has only one pair of Ag-NORs located in the subcentromeric region of the acrocentric (t) chromosome, *e.g.*, *E. guaza*, *E. alexandrinus*, *E. caninus*, *E. fasciatomaculatus*, *E. fasciatus*, and *E. awoara*; type II has one pair of Ag-NORs located in the subcentromeric region of the t chromosome pair and an extra pair of smaller Ag-NORs located on another pair of chromosomes, as in *E. adscensionis*, *E. marginatus*, and *E. malabaricus*; and type III has only one pair of Ag-NORs located on the short arm of bi-armed chromosomes, *e.g.*, *E. guttatus* and *E. coioides*. Thus, based on the available cytogenetic data on the genus *Epinephelus*, most of the NORs of groupers are located on the 24th pair of chromosome (type I), and these results are consistent with those of [18]. In this study, *E. fuscoguttatus*, *E. tukula*, and *E. lanceolatus* also belonged to type III. It is generally believed that the appearance of one pair of Ag-NORs is the ancestral character of Serranidae fish [28]. However, when Ref. [12] classified this character and compared it to data of molecular phylogenies, results were found to be irrelevant. The authors believe that the contradiction between cytogenetic and molecular phylogenetic analyses may merely be the result of insufficient data.

Hybrid breeding often produces heterosis offspring, such as offspring with a fast growth rate, strong disease resistance, or diverse morphology. For example, Liu et al. crossed different carps to obtain hybrids with a high growth rate [35]. However, many studies have found that the success possibility and whether the offspring are fertile are related to the parental karyotypes. The parents having more-similar karyotypes can increase the success ratio of hybridization [36]. At present, completely cultured groupers mainly consist of *E. akaara, E. areolatus, E. awoara, E. bleekeri, E. bruneus, E. fuscoguttatus, E. lanceolatus, E. septemfasciatus, E. tauvina, E. coioides*, and *E. malabaricus*. Establishment of karyotypic data of these groupers can provide references for crossing strategies on farms. The genetic relationship and chromosome composition of hybrid progeny can also be confirmed by a karyotype test.

Species names of different groupers have always been confusing. Most groupers living coral reef areas have similar external morphologies, and their color characteristics also may change along with their living environment. Some larvae and juveniles may even have completely different color distributions from adults, such as *E. lanceolatus* which has three irregular black spots and a brilliant color as juveniles, but becomes dark brown as adults. Therefore, identifying groupers is often controversial [1, 4, 5]. For example, *E. coioides* and *E. tauvina* are very similar and difficult to distinguish in Taiwanese waters [37]. There is still much dispute over the taxonomy of groupers when using traditional morphology. Cyt *b* gene marker is of great help in identifying similar groupers or unidentifiable fry. In the future, this marker can also be used in aquaculture breeding to reduce failures and losses with artificial reproduction.

In this study, the results showed that different groupers can be identified by analyzing the Cyt *b* gene. The phylogenetic tree constructed from the Cyt *b* gene can distinguish *Epinephelus* groupers from those in the genus *Plectropomus*. However, groupers evolved as monophyletic group, the genus *Plectropomus* is a relatively primitive group in Epinephelinae.

Epinephelus lanceolatus was previously classified in the genus *Promicrops* by [38, 39], but [6] used Cyt *b* to study molecular phylogenetic relationships of six out of 28 genera in the Serranidae, suggested that *Promicrops lanceolatus* should be classified into *Epinephelus*. Phylogenetic trees constructed with the NJ and ML methods also revealed that *E. lanceolatus* has a close relationship with other *Epinephelus* groupers [6]. In addition, scientific names of seven farmed groupers have been identified to reduce confusion and controversy.

5. Conclusions

All chromosome numbers from seven groupers (*Plectropomus leopardus, Epinephelus coioides, E. flavocaeruleus, E. fuscoguttatus, E. lanceolatus, E. polyphekadion*, and *E. Tukula*) showed a common synapomorphic character of chromosomal number, 2n = 48. Four groupers, *E. coioides, E. polyphekadion, E. fuscoguttatus*, and *E. tukula* shared the same karyotype formula of 2 sm + 46 t. *E. coioides, E. fuscoguttatus*, and *E. tukula* had one pair of Ag-NORs located on the short arm of the sm chromosome. The mitochondrial Cyt *b* gene was used to analyze phylogenetic relationships among these species. We discovered that *Epinephelus* groupers should be classified as monophyly. The minimum genetic distance expressed between *E. coioides* and *E. tukula* was 0.1276. Results showed that *E. coioides* and *E. tukula* have similar genetic characters and cell karyotypes, and should be foremost considered for artificial hybridization strategies.

Information on karyotypes of species within the *Epinephelus* is still insufficient, and further elucidation of karyotypes of *Epinephelus* will be a great help to future genetic breeding research.

Acknowledgements

The authors express their gratitude to IW Shih for assistance with laboratory work, and also thank Professor TB Yen for his help on image process and data integration.

Research funding was provided to MC Tseng by the Ministry of Science and Technology, Taiwan (NSC 102-2313-B-020-002).

Conflict of interest

Both authors, Mei-Chen Tseng and Kuan-Wei Shih declare that they have no conflict of interest.

Author details

Mei-Chen Tseng* and Kuan-Wei Shih

*Address all correspondence to: mctseng@mail.npust.edu.tw

Department of Aquaculture, National Pingtung University of Science and Technology, Pingtung, Taiwan, R.O.C.

References

[1] Heemstra PC, Randall JE. Groupers of the world (Family Serranidae, subfamily Epinephelinae). An annotated and illustrated catalogue of the grouper, rockcod, hind, coral grouper and lyretail species known to date. FAO Fisheries Synopsis. 1993;**125**:1-382

[2] Wo RMS. *Epinephelus*. In: Bailly N, editor. Fish Base. World Register of Marine Species. [Accessed: 4 Apr 2012]. http://www.marinespecies.org/aphia.php?p=taxdetails&id=126068

[3] Shao KT. Taiwan Fish Database. WWW Wed electronic publication. Version 2009/1. 2010-7-26. http://fishdb.sinica.edu.tw

[4] Randall JE, Hoese DF, Last P. On the status of the Australian serranid fishes *Epinephelus ergastularius* Whitley and *E. thompsoni* Whitley. Records of the Australian Museum. 1993; **45**:25-41

[5] Craig MT, Pondella DJII, Franck JPC, et al. On the status of the Serranid fish genus *Epinephelus*: Evidence for paraphyly based upon 16S rDNA sequence. Molecular Phylogenetics and Evolution. 2001;**19**:121-130

[6] Ding S, Zhuang X, Guo F, et al. Molecular phylogenetic relationships of China Seas groupers based on cytochrome *b* gene fragment sequences. Science in China. Series C, Life Sciences. 2006;**49**:235-242

[7] Smith CL. A revision of the American grouper: *Epinephelus* and allied genera. Bulletin of the American Museum of Natural History. 1971;**146**:241

[8] Randall JE, Ben-Tuvia A. A review of the groupers (Pisces: Serranidae: Epinephelinae) of the Red Sea, with description of a new species of *Cephalopholis*. Bulletin of Marine Science. 1983;**33**:373-426

[9] Randall JE, Heemstra PC. Revision of Indo-Pacific groupers (Perciformes: Serranidae: Epinephelinae), with descriptions of five new species. Indo-Pacific Fishes. 1991;**20**:1-296

[10] Ren XH, Cui JX, Yu QX. Chromosomal nucleolar organizer regions differentiations in Chinese cyprinid fishes. Journal of Wuhan University (Natural Science Edition). 1996;**42**: 475-480

[11] Galetti PM Jr, Molina WF, Molina PRAM, Aquilar CT. Assessing genetic diversity of Brazilian reef fishes by chromosomal and DNA markers. Genetica. 2006;**126**:161-177

[12] Wang S, Su Y, Ding S, Cai Y, Wang J. Cytogenetic analysis of orange-spotted grouper, *Epinephelus coioides*, using chromosome banding and fluorescence *in situ* hybridization. Hydrobiologia. 2010;**638**:1-10

[13] Hong M, Yang J. Studies on the Karyotype of *Epinephelus awoara*. Journal of Xiamen University (Natural Science). 1988;**27**:714-715

[14] Medrano L, Bernardi G, Couturier J, et al. Chromosome banding and genome compartmentalization in fishes. Chromosoma. 1988;**96**:178-183

[15] Martinez G, Thode G, Alvarez MC, Lopez JR. C-banding and Ag-NOR reveal heterogeneity among karyotypes of Serranids (Perciformes). Cytobios. 1989;**58**:53-60

[16] Rodríguez-Daga R, Amores A, Thode G. Karyotype and nucleolus organizer regions in *Epinephelus caninus* (Pisces, Serranidae). Caryologia. 1993;**46**:71-76

[17] Aguilar CT, Galetti PM Jr. Chromosomal studies in south atlantic Serranids (Pisces, Perciformes). Cytobios. 1997;**89**:105-114

[18] Sola L, De Innocentiis S, Gornung E, et al. Cytogenetic analysis of *Epinephelus marginatus* (Pisces: Serranidae), with the chromosome localization of the 18S and 5S rRNA genes and of the (TTAGGG) telomeric sequence. Marine Biology. 2000;**137**:47-51

[19] Ledley RS, Lubs HA, Ruddle FH. Introduction to chromosome analysis. Computers in Biology and Medicine. 1972;**2**:107-128

[20] Levan A, Fredga K, Sandberg A. Nomenclature for centromeric position on chromosomes. Hereditas. 1964;**52**:201-220

[21] Thompson JD, Higgins DG, Gibson TJ. CLUSTALW: Improving the sensitivity of progressive sequence alignment through sequence weighting, position-specific gap penalties and weight matrix choice. Nucleic Acids Research. 1994;**22**:4673-4680

[22] Kimura M. A simple method for estimating evolutionary rate of base substitution through comparative studies of nucleotide sequences. Journal of Molecular Ecology. 1980;**16**:111-120

[23] Tamura K, Stecher G, Peterson D, Filipski A, Kumar S. MEGA6: Molecular evolutionary genetics analysis version 6.0. Molecular Biology and Evolution. 2013;**30**:2725-2729. DOI: 10.1093/molbev/mst197

[24] Posada D, Buckley TR. Model selection and model averaging in phylogenetics: Advantages of Akaike information criterion and Bayesian approaches over likelihood ratio tests. Systematic Biology. 2004;**53**:793-808

[25] Nei M, Kumar S. Molecular Evolution and Phylogenetics. Oxford, UK: Oxford University Press; 2000

[26] Cho A. Constructing Phylogenetic Trees Using Maximum Likelihood. Scripps Senior Theses. 2012. 46 p. http://scholarship.claremont.edu/scripps_theses/46

[27] Felsenstein J. Confidence limits on phylogenies: An approach using the bootstrap. Evolution. 1985;**39**:783-791

[28] Molina WF, Maia-Lima FA, Affonso PRAM. Divergence between karyotypical pattern and speciation events in Serranidae fish (Perciformes). Caryologia. 2002;**55**:299-305

[29] King M. Species Evolution: The Role of Chromosomes Change. Cambridge, UK: Cambridge University Press; 1993

[30] Affonso PR, Galetti PM Jr. Chromosomal diversification of reef fishes from genus *Centropyge* (Perciformes, Pomacanthidae). Genetica. 2005;**123**:227-233

[31] Mizoguchi SMHK, Portela-Castro ALB, Martins-Santos IC. Cytogenetic characterization of *Crenicichla* (Pisces, Perciformes, Cichlidae) of the Iguaçu River. Genetics and Molecular Research. 2007;**6**:650-656

[32] Takai A, Izutsu H. Diversified chromosomal characteristics in *Centropyge* fishes (Pomacanthidae, Perciformes). Hydrobiologia. 2008;**603**:15-23

[33] Maggio T, Andaloro F, Hemida F, Arculeo M. A molecular analysis of some Eastern atlantic grouper from the *Epinephelus* and *Mycteroperca* genus. Journal of Experimental Marine Biology and Ecology. 2005;**321**:83-92

[34] Craig MT, Hastings PA. A molecular phylogeny of the groupers of the subfamily Epinephelini. Ichthyological Research. 2007;**54**:1-17

[35] Liu S, Liu Y, Zhou G, et al. The formation of tetraploid stocks of red crucian carp × common carp hybrids as an effect of interspecific hybridization. Aquaculture. 2001;**192**:171-186

[36] Zan R, Song Z. Analysis and comparison between the karyotypes of *Cyprinus carpio* and *Carassius auratus* as well as *Aristichthys nobilis* and *Hypophthalmichthys molitrix*. Acta Genetica Sinica. 1980;**7**:72-77

[37] Shen SC. Fishes of Taiwan. Department of Zoology. Taipei, Taiwan: National Taiwan University; 1993 (in Chinese)

[38] Meng QW, Su JX, Miao XZ. Fish Taxology (in Chinese). Beijing, China: China Agriculture Press; 1995

[39] Cheng QT, Zheng BS. Systematic Synopsis of Chinese Fishes. Beijing, China: Science Press; 1987 (in Chinese)

[40] Wang YX, Wang HD, Zhang HF, Liu-Fu YZ. Karyotypes of *Epinephelus coioides* and *Epinephelus akaara*. Journal of Zhanjiang Ocean University. 2004;**24**:4-8

[41] Natarajan R, Subrahmanyan K. A karyotype study of some teleosts from Portonovo waters. Proceedings of the Indiana Academy of Sciences. 1974;**79**:173-196

[42] Zheng L, Liu CW, Li CL. Studies on the karyotype of four groupers. Marine Sciences. 2005; **29**:51-55

[43] Li XQ, Peng YD. Studies on karyotype of *Epinephelus fasciatomaculosus* and *Epinephelus fasciatus*. Journal of Zhanjiang Fisheries College. 1994;**14**:22-26

[44] Zou JX, Yu QX, Zhou F. The karyotypes C-bands patterns and Ag-NORs of *Epinephelus malabaricus*. Journal of Fisheries of China. 2005;**29**:33-37

[45] Guo F, Wang J, Su YQ, et al. Study on the karyotype of *Epinephelus moara*. Marine Sciences. 2006;**8**:1-3

[46] Guo M, Wang S, Su Y, Zhou Y, Liu M, Wang J. Molecular cytogenetic analyses of *Epinephelus bruneus* and *Epinephelus moara* (Perciformes, Epinephelidae). Peer J. 2014;**2**: e412. DOI: 10.7717/peerj.412

[47] Chen Y, Rong S, Liu S, et al. Analysis of the karyotype of *Epinephelus sexfasciatus*. Journal of Zhanjiang Fisheries College. 1990;**2**:62-68

[48] Magtoon W, Donsakul T. Karyotype of five teleostean fishes from Thailand. In: Technology for global challenges. 34th Cong. Sci. Tech. Thailand, BO113; 2008

Chapter 4

Cytogenetic Tools to Study the Biodiversity of Neotropical Fish: From the Classic to the Advent of Cell Culture

Fabilene G. Paim, Maria Lígia M. de Oliveira Nobile,
Fausto Foresti and Claudio Oliveira

Additional information is available at the end of the chapter

http://dx.doi.org/10.5772/intechopen.80332

Abstract

Neotropical Ichthyofauna is considered the richest and most diverse in the world. All this biodiversity has attracted attention from researchers from different areas of study, including the cytogenetics. Many cytogenetics studies have search to understand the evolution of macro and micro karyotype structure of these different groups of fish, and classical and molecular cytogenetics techniques have contributed significantly for all knowledge of this karyotypic diversity. Recently, the use of cell cultures as an alternative to obtaining mitotic chromosomes opening up new opportunities to study groups that have not been explored or have not yet been cytogenetically investigated. In this work, we take a chronological overview of the advances of different cytogenetic techniques ("*in vivo*" and "*in vitro*" methods to obtain the chromosome, C-banding, the detection of nucleolar organizer regions (Ag-RON), fluorescent *in situ* hybridization (FISH) with several repetitive probes and paint chromosome) over the decades and how these techniques helped elucidate questions of the organization and function of the fish genome.

Keywords: chromosome, karyotype evolution, molecular cytogenetics, fibroblast cells

1. Introduction

The Neotropical region includes the area between the north of Mexico and the south of South America. This is the richest and more diversity freshwater fish fauna in the world with approximately 5160 freshwater fish species, distributed in 739 genera, 69 families and 20 orders, which

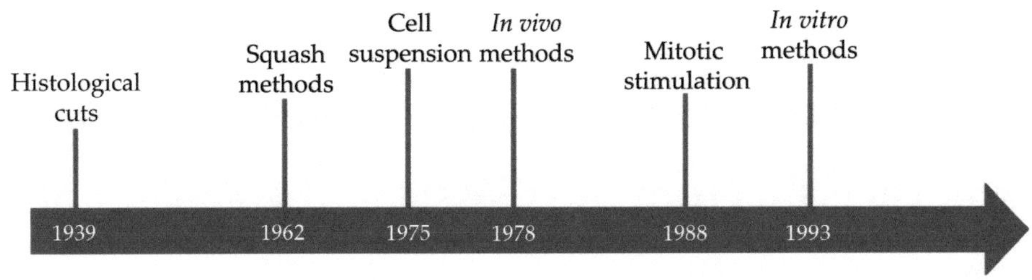

Figure 1. Timeline showing the major technical innovations that contribute to the development of fish cytogenetic.

represents one-third of all fishes on the planet [1]. A larger part of this diversity is grouped in Characiformes and Siluriformes, but there are still gaps in information in many groups [2].

All this diversity has been studied in different areas, including Cytogenetics. The refinement of cytogenetic techniques (**Figure 1**) provided the obtain of quality chromosome preparations that significantly increased the number of species studied and the description of their chromosomal characteristics, which contributed significantly to a better understanding of the genetic structures, evolution and systematic of the fishes [3–7].

2. Classical methods of cytogenetics to obtain metaphase chromosomes and their adaptations

Obtaining metaphase chromosomes is the most important point for cytogenetic studies, since any study to understand evolution and structure of the karyotypes of the species depends on this initial stage. It is known that many adjustments were made to improve the different techniques which it have arisen over the years in research within the fish cytogenetics.

The first studies with fish cytogenetics used fragments of testis previously fixed, included in paraffin and then submitted to cuts, like the experiments of Makino [8]. This methodology generated a certain doubt in its results, due to the uncertainty of the exact diploid number of each cell and was not employed with a significant number of species.

Subsequently, studies where the obtainment of chromosome depended on the squash technique were developed. In this case, a small fragment of tissue was directly crushed on a glass slide and fixed with acetic acid [9]. This technique often produced overlapping chromosomes, making it difficult to visualize the morphology and diploid number. Anyway, researches using this methodology continued and resulted in the creation of the "crushing machine" invented by Orlando Moreira Filho to minimize the injuries in the researcher's fingers [10].

The use of tissues to obtain metaphase chromosomes was not considered easily applied, because it was not easy to develop studies in the field [11]. Another relevant point is that it was not possible to regulate the rate of mitotic division and the condensation of the chromosomal arms. However, if it was known about the high hematopoietic activity of the anterior kidney in fishes [12] and from this organ, it was possible to obtain good metaphases, especially when subcutaneous or intraperitoneal stimulation of a mitogenic agent was performed [13, 14].

In 1971, Cole and Leavens [15] were the first to suggest the use of yeast as mitotic stimulating agents in hematopoietic tissues of reptiles and amphibians, but Lee and Elder [16] adapted this protocol for small mammals using a suspension of bread yeasts injected into the animal, observing that the chromosomes spread better and responded more effectively to banding treatments. For fishes, this methodology was adapted by Oliveira et al. [17], and it has been widely used over the years [18–20]. Other mitogenic agents were also employed in work with freshwater fish, such as phytohemagglutinin [21, 22], horse serum [23], parasitic infection as *Ichthyophthirius multifiliis*, or pharmaceutical agents [24, 25]. However, the use of enriched glucose solution of *Saccharomyces cerevisiae* (yeast activated suspension) is still the most used by its efficiency and low cost.

Since 1956, the works of Tjio, Levan, Ford, and Hamerton [26, 27] have reported about treating cytogenetic preparations with colchicine and hypotonic solutions, and the chromosomes have shown morphologically well-defined and that spread easily on the glass slide. Only in 1975, with the publication of a paper by McPhail and Jones, the chromosome preparations for cytogenetic studies in fish began to use this methodology [10].

The advances of chromosomal preparations in fishes have been boosted from the "air-drying" technique developed for mammals and later adapted for fish in *"in vivo"* [28] and *"in vitro"* [29] protocols. Both methods involved pre-treatment with colchicine. The use of this drug has enabled a direct control of chromosome condensation, which favored a more detailed study of the morphology of the chromosomes.

Another aspect that contributed to improving the quality of chromosome preparations was hypotonization process. Substances such as sodium citrate, distilled water and potassium chloride are used in the hypotonic treatment of the material; however, the potassium chloride is the most used in fish. In addition, the incubation temperature and hypotonization time should be adjusted according to the organism (e.g., in freshwater fish is common the hypotonization time of 21 min, while for marine fish is used from 30–35 min). After hypotonization, the cells are fixed in Carnoy's solution [21] and the cell suspension obtained is dropped into a glass slide for the rupture of the nuclear envelope [11] and thereby spreading the chromosomes for visualization of the diploid number and morphology.

Alternative methodologies have been published to improve chromosomal preparations in fish. Such methodologies, as proposed by Netto et al. [30], describe new proposed methodologies based on previously published protocols that allow cytogenetic analysis in individuals after death or that described by Blanco et al. [31], who proposed a protocol to be conducting in the field, where it eliminates the need for transportation of the specimens to the laboratory, but it is still not as common as the methodologies of Bertollo et al. [28] and Foresti et al. [29].

3. Chromosomal banding techniques and their contributions to the understanding of karyotypic macrostructure in several fish groups

Major breakthroughs in cytogenetic fish were possible with the development of differential staining techniques in the early 1970s that made it possible to understand the evolutionary relationships in many fish groups. These methods allowed a better characterization of the chromosomal structure of the fish with appearance of markings along the chromosomes that before these techniques were based only in description of the number and chromosomal morphology. The main techniques used for chromosomal characterization in fish include the C and G banding techniques (not so usual due to compartmentalization of genomics) and silver nitrate staining.

The C-banding technique described by Summer [32] shows the patterns of the constitutive heterochromatin, and it has been widely used in cytogenetic studies of fish for characterization of similar karyotypes, especially to identify variations among species or populations of the same species [19, 33–36]. It was applied for first time in salmonid species [37, 38]. In fishes from the Neotropical region, the first studies were conducted in *Prochilodus* [39], *Eigenmannia* [40], and *Leporinus* [41], and since then, several studies have reported C bands in different fish species.

Most of this heterochromatin has been reported in centromeric and terminal regions of the chromosomes of most Neotropical fish species [34, 36, 42], while in some Loricariidae species it is possible to observe many heterochromatin blocks in the interstitial region [35], which appear to be a common feature for this group. In some species, heterochromatin can be more abundant [35, 43], whereas in other species these heterochromatin blocks are reduced [42]. Other studies have emphasized the importance of heterochromatin as a major source of karyotype diversification within and among some fish groups (e.g. 19). In some groups, it is possible to observe trends in relation to the behavior of heterochromatin, for example in Hypostominae, in which there is a relationship between the amount of heterochromatin and chromosome number of the species of this subfamily [44].

Not only did the C-banding technique provided a better characterization of the karyotypes but also the use of the silver nitrate staining technique that identifies the nucleolar organizing regions (NORs) became routine since the 1980s [45]. The NORs are chromosomal regions where the ribosomal RNA genes (45S = 18S + 5.8S + 28S) are located [46]. The first works using the technique in Neotropical fishes were in the species of Gymnotiformes [40, 47].

In general, two distribution patterns of NORs can be observed in fish, the first being the occurrence of only a single chromosome pair with NORs [33, 48, 49], while in other groups of fish more sites with NORs distributed in different chromosomes of the karyotype [36, 50, 51]. In fact, a single pair of NOR has been arbitrarily considered a plesiomorphic condition in fish [52]. Although this technique has been widely used, for the cost and ease, only 1.3% of the fish species had their NORs distribution investigated [46]. In some fish groups, it has been considered an excellent cytotaxonomic marker, as in *Apareidon* and *Paradon* of the Parondontidae family [53, 54]. In addition, polymorphism NORs have been evidenced with variation and size differences between homologous NORs [47].

4. How fluorescent *in situ* hybridization (FISH) and its variations have helped in the understanding of the evolution and organization of the fish genome?

The technique of fluorescence *in situ* hybridization (FISH) made it possible to physically map specific nucleotide sequences in the chromosomes of the species or group in study [55]. It was first used by Buongiorno-Nardeli and Amaldi [56] in histological cuts and by Gall and Pardue [57] in chromosomes, but the adjustments to the protocols used to this day for fish studies are basically small changes from the original protocol proposed by Pinkel et al. [58]. This technique provided better results to investigate how chromosome diversity and organization of genomic segments occurred in fish chromosomes [59].

For example, cichlids are an interesting group of fish to be studied to explore different ecological niches and to report varied life strategies, morphology and behavior [60, 61], besides species important for fishing and aquarism [62]. Thus, many studies have search to understand more about the karyotypic macrostructure of this group of fish [66, 67], and the physical mapping of repetitive sequences has showing that such portions of the DNA may be involved in several chromosomal rearrangements in Cichlinae [63–67].

4.1. Ribosomal genes

In the genome of the eukaryotes, ribosomal genes are organized into two multigenic families, the 45S rDNA responsible for encoding the 18S, 5.8S and 26S/28S rRNAs and the other 5S rDNA, which encodes the 5S rRNA [59]. They are repeating sequences in *tandem*, and these genes are easily identified by FISH [68]. Several studies have searched to understand a little of the evolutionary dynamics of these repetitive sequences in the fish genome [5, 64, 67].

The 5S rRNA gene has been described in many fish groups, and it is located mainly in the interstitial region of the chromosome [59, 69–72], which may not only be a coincidence, but rather that this ribosomal minor distribution brings some advantage to the carrier genome [73]. It is known that the 5S rRNA is composed of a conserved region of 120 base pairs, separated from each other by the NTSs (not transcribed portions, which may vary in size or sequence). These variations have become important markers for specific species or specific populations.

Some studies with physical mapping of 5S rRNA in Anostomidae species have shown that the sites marked by the smaller ribosomal have been conserved during the karyotype evolution of the fish of this family [59, 70, 71, 74]. In *Brycon*, the physical mapping of 5S rRNA sequences was considered an important cytogenetic marker in the evolution of this group [75]. There is a variation in the number of chromosomes marked with the 5S rRNA in the genus *Astyanax*, with species with 1 pair [76, 77], species with 2 pairs [35, 76], until populations with 4 pairs, as in *A. scabripinnis* [78], and the distribution of these repetitive clusters seems to have been conserved in the group [76, 79, 80]. *Characidium* also have differences in relation to the number and location of the 5S rRNA clusters [81–83], and these variations are probably a reflection of the allopatric speciation occurred in populations of this genus.

In some fish species, more than one class of 5S rRNA gene has been identified, as reported in *Leporinus* [59]. This variation was due to differences in the sequences of portions not transcribed, and also it was reported in *Oreochromis niloticus* [84]. These sequences were found in pseudogenes and the 5S rRNA gene inverted; but in both works, the technique of FISH was contributed to identify the chromosomal location of the two classes of 5S rRNA. In the species, *Gymnotus sylvius* and *G. inaequilabiatus* were also detected two smaller classes, and with FISH, it was possible to observe that the two clusters of rRNA 5S are co-located in a chromosome pair, while the second class showed too marked in distinct chromosomes [85].

Many species of fish have the 18S rRNA gene co-located with the 5S rRNA gene [76, 86–89]; however, from the functional point of view, it would be more advantageous for two ribosomal classes to be on separate chromosomes since the transcription of them is made by distinct RNA polymerases, and the non-synteny is a way of ensuring that the 5S rRNA is not translocated to the rRNA 45S [70, 71], and allows the independent evolution of these genes [71].

Almeida-Toledo et al. [76] found that the genes 5S and 18S rRNA are co-located in five species of *Astyanax*, and such sequence was considered important markers for studying the evolutionary history of the group, including *A. altiparanae* and *A. lacustris*. This fact can be a sign of the recent separation of species, which previously belonged to a taxonomic unit of *A. bimaculatus* [90].

In the family Loricariidae, the FISH showed that most species have ribosomal sites in distinct chromosomes [91–95]. However, in the subfamily Neoplecostominae and Hypoptopomatinae [95], Hypostominae [92] and Loricariinae [91], these genes are in synteny condition, which is considered a primitive condition for the family, since it was found in the outgroup Trichomycteridae [19, 95]. According to Oliveira [19], the co-localization of 5S rDNA and 18S sites in Trichomycterus species is considered a plesiomorphic condition of the group, however the smaller ribosomal is more variable, since more labeled chromosomal pairs were observed, whereas the larger ribosomal was kept in only a couple, which according to the authors are homeologous.

Investigations using the genes rRNA 5S and 18S rRNA by Scacchetti et al. [83] showed that these genes are present in the sex chromosomes of some species of *Characidium*, indicating that the ribosomal can also participate in the differentiation process by chromosomes linked to sex in this group of fishes. In some fish, genome sequences of 18S rRNA 28S associated to heterochromatin have also been reported [69, 86], which seems to indicate that the constitutive heterochromatin may be involved in both the structural maintenance of the nucleolus and integrity of repetitions of ribosomal DNA [96].

4.2. Histones

The histone genes are composed of a genetic complex of a multigenic family (H2A, H2B, H3 and H4), which can vary in number of copies and organization genome [97]. In addition, they may be configured by H1 histone or spread throughout the genome [98]. In fishes, there are still a few studies that investigated the location and organization of these sequences, but in some of these studies histones are associated with ribosomal genes [85, 98, 99], and the genes H1, H3 and H4 are grouped in species of *Astyanax* [100, 101], as well as in the case of *Synbranchus*, where H3 and H4 are associated and spread throughout the genomes, likely to transposable elements [102]. This conformation was also observed in *Orestias ascotanensis* [103], where these sequences are organized into small copies. In *Characidium alipioi* [104], the H3 and H4 genes were mapped in a single chromosomal pair, which seems to be a conservative characteristic of the group [105].

4.3. snRNA

SnRNA genes are characterized in five RNA types (U1, U2, U4, U5, U6), non-coding, that are part of a large RNA-protein complex known as spliceosome machinery [106, 107]. The U2 gene is highly conserved in the genome of eukaryotes; however, the number of sites of these sequences may be different among species. This is because multigenic families may adopt different conservation strategies for their sequences [108].

Merlo et al. [109] and Úbeda-Manzanaro et al. [110] investigated the location of rRNA sequences U2 in species of the families Batrachoididae and Moronidae, while Manchado et al. [111] described U1 sites linked to smaller ribosomal in the genome of *Solea senegalensis*. However, few studies have been performed to map these sequences in Neotropical fishes. Study conducted by Cabral-de-Melo et al. [112] showed that the U1 snRNA gene in cichlids is found in just one chromosome pair, probably being a conserved feature in this group since the fragmentation of Gondwana [113]. On the other hand, the technique of FISH showed that the position of the snRNA U1 clusters can vary between distant species, and this is due to chromosomal rearrangements such as inversions and transpositions that modify and restructure the karyotypes of cichlids. The snRNA U1 sites were more variable between South American Cichlids than among the African species [112].

In *Gymnotus*, physical mapping of U2 snRNA sequences showed differences in the distribution of this gene, which can be clustered in homologous chromosomes as in most species or spread in several sites as in *G. pantanal*, an apomorphic condition [102]. In addition, the technique of FISH showed the U2 snRNA marked in a chromosome linked to sex in the species *G. pantanal* [102]. In other Neotropical fishes, these two configurations of the location of U2 snRNA gene can be found [83, 102, 103, 113, 114].

4.4. Telomeric probes

The telomere portions of the chromosomes are composed of repetitive sequences in tandem, which in vertebrates have been reported by sequence (TTAGGG)n [115]. In fish, these sites have already been mapped occupying regions of the telomeres [116, 117] and non-telomeric chromosome portions [118]. These interstitial marks contribute to studies about organization and macrostructural evolution of karyotypes, since they may answer some questions as fusion or inversions that modify the chromosomal structure of some species [117, 119]. Sometimes, these interstitial sites are the result of fusions but are not easily mapped because the karyotype in study may evolve and the telomeric sequences lose its function [55]. Another relevant point investigated in fish with a FISH technique using telomeric sequences is associated with satellite DNA [120, 121], which would be a response to the spreading of these regions in the interstitial regions. Scacchetti et al. [121] made it through the physical mapping of telomeric sequences in *Characidium* species, find interstitial markings in the chromosomes of some populations and, from there, carried out analyses that allowed establishing monophyletic group conditions. In Cioffi and Bertollo [122], telomeric interstitial markings were also observed in the neo-Y genome chromosome *Hoplias malabaricus*, which contributed to answer questions about the origins of the sexual system in this group of fishes.

4.5. Satellite DNA

Satellite DNA is composed of repetitive sequences that tend to accumulate in the chromosomes, especially in heterochromatic regions [123]. They are not protein coding and can form clusters on the chromosome arms [123], which facilitate their physical mapping in the karyotype of interest. In the 1980s, satellite DNA families were first described in fish, and many works showed that they accumulate in the centromeric portion of the chromosomes and they may be related to the structural and functional roles of the centromere [124–126]. Some events such as unequal crossing over, transpositions and duplications may contribute to repetitive sequences including satellite DNA accumulating in heterochromatic regions, where they undergo less selective pressures and may thus evolve in the genome [127]. Some studies have used different satellite DNA probes to investigate the composition of supernumerary chromosomes in some species of fish [128–131].

4.6. Sex chromosome

Several studies attempted to understand the origin, evolution and maintenance of the sex-linked chromosomes [103], and fishes have become excellent models of studies because they have a wide and varied sexual system [122]. The sex chromosomes have been described in more than 7% of the fish karyotypes [132], and with the FISH technique, many satellite DNA sequences have been isolated and mapped in different species [133–137]. In some species of fishes, FISH technique has contributed to map sequences that characterize sex chromosomes undifferentiated by morphology or conventional staining, as is the case with guppy, within the family Poeciliidae [133, 138, 139]. And in other cases, the mappings of satellites sequences were important in work with morphologically differentiated sex chromosomes [135, 140]. Chromosome painting using W-chromosome-specific probe helped to answer about the common origin of this chromosome linked to sex in *Characidium* species [83, 141].

4.7. B chromosome

Many studies search to understand more about the origin, function and evolution of B chromosomes in fishes, since these are considered expendable parasites to supernumerary genome [142]. With the technique of FISH and advances in chromosomal painting, studies using themselves as probes it was possible to examine if there is homology of these extra chromosomes with the normal chromosomes of the karyotype, and from this understand possible answers about the origin and evolution of these chromosomes [51, 104, 130, 143, 144].

4.8. Fiber-FISH

The Fiber-FISH technique contributed greatly to the investigation of specific sites in the genome of Neotropical fish, since it allowed to determine the position of the genes in the chromatin fiber and to verify the organization of the gene sequences [145].

5. Culture of cells in fish: alternative tools for obtaining metaphase chromosomes

Cell culture is an *in vitro* technique widely used to isolate and maintain cells outside their original environment [146]. Briefly, a tissue fragment is aseptically removed from the individual and then mechanically and enzymatically dissociated or both. The isolated cells are

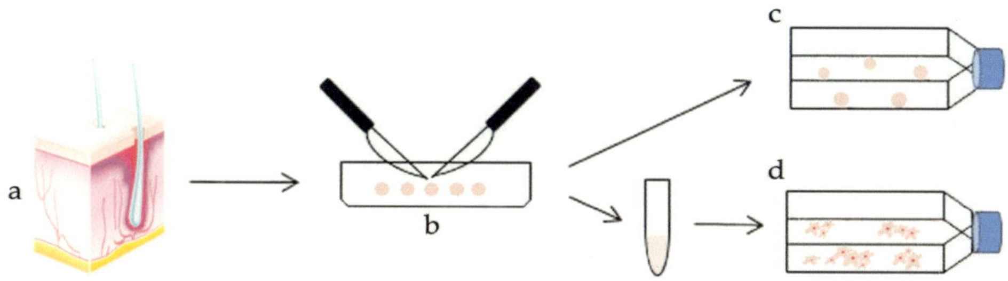

Figure 2. Scheme of obtaining cell culture. (a) Tissue; (b) disaggregation of the tissue mechanically; (c) tissue fragments (explant) grown in flasks with medium culture and (d) cells cultured in flasks after enzymatic disaggregation.

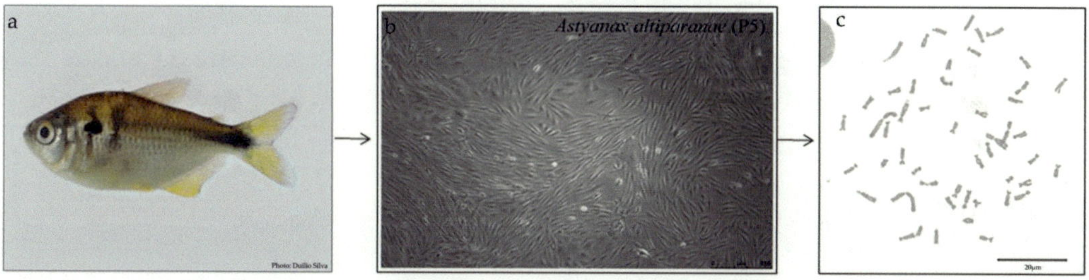

Figure 3. (a) *Astyanax altiparanae* (Characidae); (b) fibroblast cell line of *A. altiparanae* in the fifth passages and (c) mitotic chromosome of *A. altiparanae* obtain a cell line with diploid number of 2n = 50 chromosomes.

cultured in flasks with suitable medium with adjusted pH, antibiotic/antimycotic agents and fetal bovine serum (**Figure 2**). Cell culture is maintained at the appropriate temperature to the species under study and monitored daily for cell growth and possible contamination. When these cells cover the entire bottom of the flask (cell confluence), these cells are trypsinized and cultured in new vials (a process known as subculture or passage). These cells are treated with colchicine and after detached from the bottom of the flask are hypotonized, fixed with Carnoy's solution and then dropped onto slides (**Figure 3**) [147, 148].

Cell culture is still little used as a tool in fish cytogenetic studies [149–151], mainly by the difficulty of standardization of the technique of isolation and maintenance of cell cultures. Nevertheless, this technology is an excellent alternative to obtain good quality chromosome preparations, since it can be applied in cytogenetic studies of small and large species, in which it is difficult to work with direct methods of chromosome preparation or also in species used in aquaculture or endangered, when there is no possibility of sacrifice of animals [149]. Another advantage is that the methodology can provide the establishment of cell bank available at any time, so, in case of repetition of cytogenetic methodologies, it is not necessary to go back to the field for new individuals.

6. Conclusion

The advances of cytogenetic techniques have contributed directly in studies that search to investigate and understand the macro and micro karyotype structure of the most diverse groups of Neotropical fish, and many questions have been answered with the use of these technologies, as well as new problems have arisen that it was not possible to investigate because of the difficulties of the techniques. It is known that there are still many gaps to be filled, but cytogenetics has grown a lot in recent years and morphological and /or phytogenetic tools have played an important role in cytogenetic advances.

Author details

Fabilene G. Paim*, Maria Lígia M. de Oliveira Nobile, Fausto Foresti and Claudio Oliveira

*Address all correspondence to: fabillene@yahoo.com.br

Departamento de Morfologia, Instituto de Biociências, Universidade Estadual Paulista 'Júlio de Mesquita Filho', Brazil

References

[1] Reis RE, Albert JS, Dario FD, Mincarone MM, Petry P, Rocha LA. Fish biodiversity and conservation in South America. Journal of Fish Biology. 2016;**89**:1-39. DOI: 10.1111/jfb.13016

[2] Albert JS, Reis RE. Historical Biogeography of Neotropical Freshwater Fishes. Berkeley: University of California Press; 2011

[3] Porto-Foresti F, Hashimoto DT, Alves AL, Almeida RBC, et al. Cytogenetic markers as diagnoses in the identification of the hybrid between piauçu (*Leporinus macrocephalus*) and piapara (*Leporinus elongatus*). Genetics and Molecular Biology. 2008;**31**(1):195-202

[4] Hashimoto DT, Laudicina A, Bortolozzi J, Foresti F, Porto-Foresti F. Chromosomal features of nucleolar dominance in hybrids between the Neotropical fish *Leporinus macrocephalus* and *Leporinus elongatus* (Characiformes, Anostomidae). Genetica. 2009;**137**:135-140. DOI: 10.1007/s10709-009-9366-y

[5] Vicari MR, Nogaroto V, Noleto RB, Cestari MM, Cioffi MB, et al. Satellite DNA and chromosomes in Neotropical fishes: Methods, applications and perspectives. Journal of Fish Biology. 2010;**76**:1094-1116. DOI: 10.1111/j.1095-8649.2010.02564.x

[6] Tenório RCCO, Vitorino CA, Souza IL, Oliveira C, Venere PC. Comparative cytogenetics in *Astyanax* (Characiformes: Characidae) with focus on the cytotaxonomy of the group. Neotropical Ichthyology. 2013;**11**(3):553-564

[7] Ramirez JL, Birindelli JLO, Galetti PM. A new genus of Anostomidae (Ostariophysi: Characiformes): Diversity, phylogeny and biogeography based on cytogenetic, molecular and morphological data. Molecular Phylogenetics and Evolution. 2017;**107**:308-323. DOI: 10.1016/j.ympev.2016.11.012

[8] Makino S. The chromosomes of the Carp, *Cyprinus caypio*, including those of some related species of Cyprinidae for comparison. Cytologia. 1939;**9**:430-440

[9] Bstergren G, Heneen WK. A Squash Technique for Chromosome Morphological Studies. Sweden: Institute of Genetics, University of Lund; 1962. pp. 332-341

[10] Foresti F. A brief history of Fish genetics in Brazil. Genetics and Molecular Biology. 2008;**31**(1 (suppl)):1-4

[11] Sessionss K. Chromosomes: Molecular cytogenetics. In: Hillis DM, Moritz C, editors. Molecular Systematics. Sunderland, Massachusetts: Sinauer Associates; 1990. pp. 156-203

[12] Catton WT. Blood cell formation in certain teleost fishes. Blood. 1951;**6**:39-60

[13] Ojima Y, Kurishita A. A new method to increase the number of mitotic cells in the kidney tissue for fish chromosome studies. Proceedings of the Japan Academy. 1967;**56** (10):610-615

[14] Ozouf-Costaz C, Foresti F. Fish cytogenetic research: Advances, applications and perspectives. Netherlands Journal of Zoology. 1992;**42**(2-3):277-290

[15] Cole CI, Leavens CR. Chromosome preparations of amphibians and reptiles: Improved techniques. Herpet. 1971;**3**(6):102

[16] Lee MR, Elder FFB. Yeast stimulation of bone marrow mitosis for cytogenetic investigations. Cytogenetics and Cell Genetics. 1980;**26**:36-40

[17] Oliveira C, Almeida-Toledo LF, Foresti F, Toledo Filho SA. Supernumerary chromosomes, Robertsonian rearrangement and multiple NORs in *Corydoras aeneus* (Pisces, Siluriformes, Callichthyidae). Caryologia. 1988;**41**(36):227-236

[18] Nirchio M, Rondón R, Oliveira C, et al. Cytogenetic studies in three species of *Lutjanus* (Perciformes: Lutjanidae: Lutjaninae) from the Isla Margarita, Venezuela. Neotropical Ichthyology. 2008;**6**(1):101-108

[19] Oliveira MLM, Utsunomia R, Pansonato-Alves JC, et al. Microstructural chromosome reorganization in the genus *Trichomycterus* (Siluriformes: Trichomycteridae). Neotropical Ichthyology. 2016;**14**(2):e150084. DOI: 10.1590/1982-0224-20150084

[20] Paim FG, Almeida LAH, Affonso PRAM, Sobrinho-Scudeler PE, et al. Chromosomal stasis in distinct families of marine Percomorpharia from South Atlantic. Comparative Cytogenetics. 2017;**11**(2):299-307. DOI: 10.3897/CompCytogen.11(2).11942

[21] Gold JR, Li YC, Shipey NS, Powers PK. Improved methods for working with fish chromosomes with a review of metaphase chromosome banding. Journal of Fish Biology. 1990;**37**:563-575

[22] Ren X, Qixing Y, Ping W. Polymorphisms of silver stained NORs in rice-fish eels. Acta Genetica Sinica. 1991;**18**:304-311

[23] Ojima Y, Kurishita A. A new method to increase the number of mitotic cells in the kidney for fish chromosome study. Proceedings of the Japan Academy, Series B. 1980;**56**:610-615

[24] Molina WF. An alternative method for mitotic stimulation in fish cytogenetics. Chromosome Science. 2001;**5**:149-152

[25] Molina WF, Alves DEO, Araújo WC, Martinez PA, et al. Performance of human immunostimulating agents in the improvement of fish cytogenetic preparations. Genetics and Molecular Research. 2010;**9**(3):1807-1814

[26] Tjio JH, Levan A. The Chromosome Number of Man. Lund: Institute of Genetics; 1956. pp. 1-6

[27] Ford CE, Hamerton JL. A colchicine, hypotonic citrate squash sequence for mammalian chromosomes. Stain Technology. 1956;**31**:247-251

[28] Bertollo LAC, Takahashi CS, Moreira-Filho O. Cytotaxonomic considerations on *Hoplias lacerdae* (Pisces, Erythrinidae). Brazilian Journal of Genetics. 1978;**2**:103-120

[29] Foresti F, Oliveira C, Almeida-Almeida-Toledo LF. A method for chromosome preparations from large fish specimens using in vitro short-term treatment with colchicine. Experientia. 1993;**49**:810-813

[30] Netto MRCB, Pauls E, Affonso PRAMA. Standard protocol for obtaining fish chromosomes under post-mortem conditions. Micron. 2007;**38**:214-217. DOI: 10.1016/j.micron.2006.07.019

[31] Blanco DR, Vicari MR, Lui RL, Artoni RF, Almeida MC, et al. Origin of the X1X1X2X2/X1X2Y sex chromosome system of *Harttia punctata* (Siluriformes, Loricariidae) inferred from chromosome painting and FISH with ribosomal DNA markers. Genetica. 2014;**142**:119-126. DOI: 10.1007/s10709-014-9759-4

[32] Sumner AT. A simple technique for demonstrating centromeric heterochromatin. Experimental Cell Research. 1972;**75**:304-306. DOI: 10.1016/0014-4827(72)90558-7

[33] Galetti PM Jr, Cesar ACG, Venere PC. Heterochromatin and NORs variability in *Leporinus* fish (Anostomidae, Characiformes). Caryologia. 1991;**44**(3-4):287-292. DOI: 10.1080/00087114.1991.10797193

[34] Balen RE, Noleto RB, Vicari MR, Artoni RF, Cestari MM. Comparative cytogenetics among populations of *Hollandichthys multifasciatus* (Teleostei: Characidae). Zoological Science. 2013;**30**(2):105-109. DOI: 10.2108/zsj.30.105

[35] Kavalco KF, Pazza R, Almeida-Toledo LF. *Astyanax bockmanni* Vari and Castro, 2007: An ambiguous karyotype in the Astyanax genus. Genetica. 2009;**136**:135-139

[36] Kamei MCSL, Baumgärtner L, Paiva S, Zawadzki CH, et al. Chromosomal diversity of three species of *Hypostomus* Lacépède, 1803 (Siluriformes, Loricariidae), from the Paraná River Basin, Brazil: A species complex in *Hypostomus ancistroides* reinforced by a ZZ/ZW sex chromosome system. Zebrafish. 2017;**14**:4. DOI: 10.1089/zeb.2017.1429

[37] Abe S, Muramoto J. Differential staining of chromosomes of two Salmonoid species, *Salvelinus leucomaenis* (Pallas) and *Salvelinus malma* (Walbaum). Proceedings of the Japan Academy. 1974;**50**:507-511

[38] Zenzes MT, Voiculescu O. C-banding of patterns in *Salmo trutta* a species of tetraploid origin. Genetica. 1975;**45**:531-536

[39] Pauls E, Bertollo LAC. Evidence for a system of supernumerary chromosomes in *Prochilodus scrofa* Steindachner, 1881 (Pisces, Prochilodontidae). Caryologia. 1983;**36**(4): 307-314. DOI: 10.1080/00087114.1983.10797671

[40] Almeida-Toledo LF, Foresti H, Toledo Filho SA. Complex sex chromosome system in *Eigenmannia* sp. (Pisces, Gymnotiformes). Genetica. 1984;**64**:165-169

[41] Galetti PM, Foresti F. Evolution of the ZZ/ZW system in *Leporinus* (Pisces, Anostomidae). Cytogenetics and Cell Genetics. 1986;**43**:43-46

[42] Artoni RF, Bertollo LAC. Evolutionary aspects of the ZZ/ZW sex chromosome system in the Characidae fish, genus *Triportheus*. A monophyletic state and NOR location on the W chromosome. Heredity. 2002;**89**:15-19. DOI: 10.1038/sj.hdy.6800081

[43] Galetti Jr PM, Mestriner CA, Venere PC, Foresti F. Heterochromatin and karyotypic reorganization in fish of family Anostomidae (Characiformes). Cytogenetics and Cell Genetics. 1991;**56**:116-121

[44] Artoni RF, Bertollo LAC. Trends in the karyotype evolution of Loricariidae fish (Siluriformes). Hereditas. 2001;**134**:201-210

[45] Howell WM, Black DA. Controlled silver staining of nucleolus organizer region with protective colloidal developer: A 1-step method. Experientia. 1980;**36**:1014-1015. DOI: 10.1007/BF01953855

[46] Gornung E. Twenty years of physical mapping of major ribosomal RNA genes across the teleosts: A review of research. Cytogenetic and Genome Research. 2013;**141**:90-102. DOI: 10.1159/000354832

[47] Foresti F, Almeida Almeida-Toledo LF, Almeida-Toledo FSA. Polymorphic nature of nucleolus organizer regions in fishes. Cytogenetics and Cell Genetics. 1981;**31**:137-144. DOI: 10.1159/000131639

[48] Oliveira C, Almeida-toledo LF, Almeida-Toledo FSA. Comparative cytogenetic analysis of three cytotypes of *Corydoras nattereri* (Pisces, Siluriformes, Callichthyidae). Cytologia. 1990;**55**:21-26

[49] Martinez ERM, Alves AL, Silveira SM, Foresti F, Oliveira C. Cytogenetic analysis in the incertae sedis species *Astyanax altiparanae* Garutti and Britzki, 2000 and *Hyphessobrycon eques* Steindachner, 1882 (Characiformes, Characidae) from the upper Paraná river basin. Comparative Cytogenetics. 2012;**6**(1):41-51. DOI: 10.3897/CompCytogen. v6i1.1873

[50] Born GG, Bertollo LAC. A new sympatric region for distinct karyotypic forms of *Hoplias malabaricus* (Pisces, Erythrinidae). Brazilian Journal of Biology. 2006;**66**(1B):205-210

[51] Silva DMZ, Pansonato-Alves JC, Utsunomia R, Araya-Jaime C, Ruiz-Ruano FJ, et al. Delimiting the origin of a B chromosome by FISH mapping, chromosome painting and

DNA sequence analysis in *Astyanax paranae* (Teleostei, Characiformes). PLoS One. 2014; **9**(4):e94896. DOI: 10.1371/journal.pone.0094896

[52] Amemiya CT, Gold JR. Chromosomal NORs as taxonomic and systematic characters in North American cyprinid fishes. Genetica. 1998;**76**:81-90

[53] Bellafronte E, Margarido VP, Moreira-Filho O. Cytotaxonomy of *Parodon nasus* and *Parodon tortuosus* (Pisces, Characiformes). A case of synonymy confirmed by cytogenetic analyses. Genetics and Molecular Biology. 2005;**28**(4):710-716

[54] Jesus CM, Moreira-Filho O. Karyotypes of three species of *Parodon* (Teleostei, Parodontidae). Ichthyological Exploration of Freshwaters. 2000;**11**:75-80

[55] Guerra M. FISH: Conceitos e aplicações na citogenética. Ribeirão Preto, SP, BR: Sociedade Brasileira de Genética; 2004. 184 pp

[56] Buongiorno-Nardelli M, Amaldi F. Autoradiographic detection of molecular hybrids between rRNA and DNA in tissue sections. Nature. 1969;**225**:946-947

[57] Gall J, Pardue ML. Formation and detection of RNA-DNA hybrid molecules in cytological preparations. Proceedings of the National Academy of Sciences. 1969;**63**:378-383

[58] Pinkel D, Straume T, Gray JW. Cytogenetic analysis using quantitative, high-sensitivity, fluorescence hybridization. Proceedings of the National Academy of Sciences. 1986;**83**:2934-2938. DOI: 10.1073/pnas.83.9.2934

[59] Martins C, Galetti PM Jr. Organization of 5S rDNA in *Leporinus* fish species: Two different genomic locations are characterized by distinct nontranscribed spacers. Genome. 2001;**44**:903-910

[60] Lowe-McConnell RH. Ecology of cichlids in South American and African waters, excluding the African Great Lakes. In: Keenleyside MHA, editor. Cichlid Fishes: Behavior, Ecology and Evolution. London: Chapman and Hall; 1991. pp. 60-85

[61] López-Fernández H, Honeycutt RL, Winemiller KO. Molecular phylogeny and evidence for an adaptive radiation of geophagine cichlids from South America (Perciformes: Labroidei). Molecular Phylogenetics and Evolution. 2005;**34**:227-244. DOI: 10.1016/j.ympev.2004.09.004

[62] Schnider CH, Gross MC, Terencio ML, Artoni RF, et al. Chromosomal evolution of neotropical cichlids: The role of repetitive DNA sequences in the organization and structure of karyotype. Reviews in Fish Biology and Fisheries. 2013;**23**:201-214. DOI: 10.1007/s11160-012-9285-3

[63] Gross MC, Schneider CH, Valente GT, et al. Comparative cytogenetic analysis of the genus *Symphysodon* (Discus Fishes, Cichlidae): Chromosomal characteristics of Retrotransposons and minor ribosomal DNA. Cytogenetic and Genome Research. 2009;**127**:43-53. DOI: 10.1159/000279443

[64] Gross MC, Schneider CH, Valente GT, et al. Variability of 18S rDNA locus among Symphysodon fishes: Chromosomal rearrangements. Journal of Fish Biology. 2010;**76**: 1117-1127. DOI: 10.1111/j.1095-8649.2010.02550.x

[65] Teixeira WG, Ferreira IA, Cabral-de-Mello DC, et al. Organization of repeated DNA elements in the genome of the cichlid fish Cichla kelberi and its contributions to the knowledge of fish genomes. Cytogenetic and Genome Research. 2009;**125**:224-234. DOI: 10.1159/000230006

[66] Mazzuchelli J, Martins C. Genomic organization of repetitive DNAs in the cichlid fish *Astronotus ocellatus*. Genetica. 2009;**136**:461-469. DOI: 10.1007/s10709-008-9346-7

[67] Valente GT, Mazzuchelli J, Ferreira IA, Poletto AB, et al. Cytogenetic mapping of the retroelements Rex1, Rex3 and Rex6 among Cichlid fish: New insights on the chromosomal distribution of transposable elements. Cytogenetic and Genome Research. 2011;**133**:34-42. DOI: 10.1159/000322888

[68] Rebordinos L, Cross I, Merlo A. High evolutionary dynamism in 5S rDNA of fish: State of the art. Cytogenetic and Genome Research. 2013;**141**:103-113

[69] Fujiwara A, Abe S, Yamaha E, Yamazaki F, Yoshida MC. Chromosomal localization and heterochromatin association of ribosomal RNA genes loci and silver stained nucleolar organizer regions in salmonid fishes. Chromosome Research. 1998;**6**:463-471

[70] Martins C, Galetti PM Jr. Chromosomal localization of 5S rDNA genes in *Leporinus* fish (Anostomidae, Characiformes). Chromosome Research. 1999;**7**:363-367

[71] Martins C, Galetti PM Jr. Conservative distribution of 5S rDNA loci in *Schizodon* (Pisces, Anostomidae) chromosomes. Chromosome Research. 2000;**8**:353-355

[72] Born GG, Bertollo LAC. An XX/XY sex chromosome system in a fish species, *Hoplias malabaricus* with a polymorphic NOR bearing X chromosome. Chromosome Research. 2000;**8**:111-118

[73] Martins C, Wasko AP. Organization and evolution of 5S ribosomal DNA in the fish genome. In: Williams CR, editor. Focus on Genome Research. Hauppauge: Nova Science Publishers; 2004. pp. 335-363

[74] Aguilar CT. Estudos citogenéticos e moleculares em populações brasileiras de Leporellus vittatus (Characiformes, Anostomidae). Rio de Janeiro: Tese (Doutorado)—Universidade Federal do Rio de Janeiro; 2001

[75] Wasko AP, Martins C, Wright JM, Galetti PM Jr. Molecular organization of 5S rDNA in fishes of the genus *Brycon*. Genome. 2001;**44**:893-902

[76] Almeida-Toledo LF, Ozouf-Costaz C, Foresti F, Bonillo C, et al. Conservation of the 5S bearing chromosome pair and co-localization with major rDNA clusters in five species of *Astyanax* (Pisces, Characidae). Cytogenetic and Genome Research. 2002;**97**:229-233

[77] Vicari MR, Artoni RF, Moreira-Filho O, Bertollo LAC. Diversification of a ZZ/ZW sex chromosome system in Characidium fish (Crenuchidae, Characiformes). Genetica. 2008;**134**:311-317. DOI: 10.1007/s10709-007-9238-2

[78] Ferro DAM, Neo DM, Moreira-Filho O, Brtollo LA. Nucleolar organizing regions, 18S and 5S rDNA in *Astyanax scabripinnis* (Pisces, Characidae): Populations distribution and functional diversity. Genetica. 2001;**110**:55-62

[79] Mantovani M. Citogenetica comparativa entre populações de Astyanax scabripinnis (Pisces, Characidae) da bacia do rio Paranapanema [Dissertação (Mestrado)]. Universidade Federal de São Carlos; 2001

[80] Hashimoto DT, Ferguson-Smith MA, Rens W, Foresti F, Porto-Foresti F. Chromosome mapping of H1 histone and 5S rRNA gene clusters in three species of *Astyanax* (Teleostei, Characiformes). Cytogenetic and Genome Research. 2011;**134**:64-71. DOI: 10.1159/000323512

[81] Pansonato-Alves JC, Oliveira C, Foresti F. Karyotypic conservatism in samples of *Characidium* cf. *zebra* (Teleostei, Characiformes, Crenuchidae): Physical mapping of ribosomal genes and natural triploidy. Genetics and Molecular Biology. 2011;**34**:208-213. DOI: 10.1590/S1415-47572011005000005

[82] Pucci MB, Barbosa P, Nogaroto V, Almeida MC, Artoni RF, Pansonato-Alves JC, et al. Population differentiation and speciation in the genus *Characidium* (Characiformes: Crenuchidae): Effects of reproductive and chromosomal barriers. Biological Journal of the Linnean Society. 2014;**111**:541-553. DOI: 10.1111/bij.12218

[83] Scacchetti PC, Utsunomia R, Pansonato-Alves JC, Da Silva GJC, et al. Repetitive DNA sequences and evolution of ZZ/ZW sex chromosomes in *Characidium* (Teleostei: Characiformes). Plos One. 2015;**10**:e0137231. DOI: 10.1371/journal.pone.0137231

[84] Martins C, Wasko AP, Oliveira C, Porto-Foresti F, Parise-Maltempi PP, et al. Dynamics of 5S rDNA in the tila (*Oreochromis niloticus*) gnome: Repeat units, inverted sequences, pseudogenes and chromosome loci. Cytogenetic and Genome Research. 2002;**98**:78-85

[85] Scacchetti PC, Alves JCP, Utsunomia R, Claro FL, de Toledo LFA, et al. Molecular characterization and physical mapping of two classes of 5S rDNA in the genomes of *Gymnotus sylvius* and *G. inaequilabiatus* (Gymnotiformes, Gymnotidae). Cytogenetic and Genome Research. 2012;**136**:131-137. https://doi.org/10.1159/000335658

[86] Pendás AM, Móran P, Freije JP, Garcia-Vásquez E. Chromosomal location and nucleotide sequence of two tandem repeats of the Atlantic salmon 5S rDNA. Cytogenetics and Cell Genetics. 1994;**67**:31-36

[87] Hatanaka T, Galetti PM Jr. Mapping of the 18S and 5S ribosomal RNA genes in the fish *Prochilodus argenteus* Agassiz, 1829 (Characiformes, Prochilodontidae). Genetica. 2004;**122**:239-244

[88] Móran P, Martínez JL, Garcia-Vásquez E, Pendás AM. Sex linkage of 5S rDNA in rainbow trout (*Oncorhynchus mykiss*). Cytogenetics and Cell Genetics. 1996;**75**:145-150

[89] Moraes-Neto A, Silva M, Matoso DA, Vicari MR, et al. Karyotype variability in neotropical catfishes of the family Pimelodidae (Teleostei: Siluriformes). Neotropical Ichthyology. 2011;**9**:97-105

[90] Garutti V, Britski HA. Descrição de uma espécie novade *Astyanax* (Teleostei: Characidae) da Bacia do Alto Rio Paraná e consideraço˜es sobre as demais espécies do gênero na bacia. Comunicações do Museu de Ciencias da PUCRS, Série Zoologia. 2000

[91] Kavalco KF, Pazza R, Bertollo LA, Moreira-Filho O. Gene mapping of 5S rDNA sites in eight fish species from the Paraiba do Sul river basin, Brazil. Cytogenetic and Genome Research. 2004;**106**:107-110

[92] Mariotto S, Centofante L, Vicari MR, Artoni RFA, et al. Chromosomal diversification in ribosomal DNA sites in *Ancistrus* Kner, 1854 (Loricariidae, Ancistrini) from three hydrographic basins of Mato Grosso, Brazil. Comparative Cytogenetics. 2011;**5**:289-300

[93] Mendes-Neto EO, Vicari MR, Artoni RF, Moreira-Filho O. Description of karyotype in *Hypostomus regani* (Ihering, 1905) (Teleostei, Loricariidae) from the Piumhi river in Brazil with comments on karyotype variation found in *Hypostomus*. Comparative Cytogenetics. 2011;**5**:133-142

[94] Rosa KO, Ziemniczak K, Barros AV, Nogaroto V, et al. Numeric and structural chromosome polymorphism in *Rineloricaria lima* (Siluriformes: Loricariidae): Fusion points carrying 5S rDNA or telomere sequence vestiges. Reviews in Fish Biology and Fisheries. 2011;**22**:739-749

[95] Ziemniczak K, Barros AV, Rosa KO, Nogaroto V, et al. Comparative cytogenetics of Loricariidae (Actinopterygii: Siluriformes): Emphasis in Neoplecostominae and Hypoptopomatinae. The Italian Journal of Zoology. 2012;**79**:1

[96] McStay B, Grummt I. The epigenetics of rRNA genes: From molecular to chromosome biology. Annual Review of Cell and Developmental Biology. 2008;**24**:131-157. DOI: 10.1146/annurev.cellbio.24.110707.175259

[97] Kedes LH. Histone genes and histone messengers. Annual Review of Biochemistry. 1979;**48**:837-870

[98] Childs G, Maxson R, Cohn RH, Kedes L. Orphons: Dispersed genetic elements derived from tandem repetitive genes of eukaryotes. Cell. 1981;**23**:651-663

[99] Lima-Filho P, Cioffi M, Bertollo L, Molina WF. Chromosomal and morphological divergences in Atlantic populations of the frillfin goby *Bathygobius soporator* (Gobiidae, Perciformes). Journal of Experimental Marine Biology and Ecology. 2012;**434-435**:63-70. DOI: 10.1016/j.jembe.2012.08.004

[100] Pansonato-Alves JC, Hilsdorf AWS, Utsunomia R, et al. Chromosomal mapping of repetitive DNA and cytochrome C oxidase I sequence analysis reveal differentiation among sympatric samples of *Astyanax fasciatus* (Characiformes, Characidae). Cytogenetic and Genome Research. 2013;**141**:133-142. DOI: 10.1159/000354885

[101] Silva DMZA, Pansonato-Alves JC, Utsunomia R, Daniel SN, et al. Chromosomal organization of repetitive DNA sequences in *Astyanax bockmanni* (Teleostei, Characiformes): Dispersive location, association and co-localization in the genome. Genetica. 2013;**141**: 329-336

[102] Utsunomia R, Pansonato-Alves JC, Scacchetti PC, Oliveira C, Foresti F. Scattered organization of the histone multigene family and transposable elements in *Synbranchus*. Genetics and Molecular Biology. 2014;**37**:30-36. DOI: 10.1590/S1415-47572014000100007

[103] Araya-Jaime C, Lam N, Pinto IV, Méndez MA, Iturra P. Chromosomal organization of four classes of repetitive DNA sequences in killifish *Orestias ascotanensis* Parenti, 1984 (Cyprinodontiformes, Cyprinodontidae). Comparative Cytogenetics. 2017;**11**(3):463-475. DOI: 10.3897/CompCytogen.v Iii3.11729

[104] Serrano EA, Utsunomia R, Scudeller PS, Oliveira C, Foresti F. Origin of B chromosomes in *Characidium alipioi* (Characiformes, Crenuchidae) and its relationships with supernumerary chromosomes in other *Characidium* species. Comparative Cytogenetics. 2017;**11**(1):81-95. DOI: 10.3897/CompCytogen.v11i1.10886

[105] Oliveira et al. Physical mapping of repeating sequences in the genome of *Characidium* and investigation of two classes of 5S rDNA in the species (in preparation)

[106] Bringmann P, Lührmann R. Purification of the individual snRNPs U1, U2, U5 and U4/U6 from HeLa cells and characterization of their protein constituents. EMBO Journal. 1986;**5**(13):3509-3516

[107] Valadkhan S. snRNAs as the catalysts of pre-mRNA splicing. Current Opinion in Chemical Biology. 2005;**9**:603-608

[108] Matera AG, Weiner AM, Schmid CW. Structure and evolution of the U2 small nuclear RNA multigene family in primates: Gene amplification under natural selection? Molecular and Cellular Biology. 1990;**10**:5876-5882

[109] Merlo MA, Cross I, Chairi H, Manchado M, Rebordinos L. Analysis of three multigene families as useful tolls in species characterization of two closely-related species, *Dicentrarchus labrax, Dicentrarchus punctatus* and their hybrids. Genes & Genetic Systems. 2010;**85**:341-349

[110] Úbeda-Manzanaro M, Merlo MA, Palazón JL, Cross I, Sarasquete C, Rebordinos L. Chromosomal mapping of the major and minor ribosomal genes, (GATA)n and U2 snRNA gene by double-colour FISH in species of the Batrachoididae family. Genetica. 2010;**138**:787-794

[111] Manchado M, Zuasti E, Cross I, Merlo A, Infante C, Rebordinos L. Molecular characterization and chromosomal mapping of the 5S rRNA gene in Solea senegalensis: A new linkage to the U1, U2, and U5 small nuclear RNA genes. Genome. 2006;**49**:79-86

[112] Cabral-de-Mello DC, Valente GT, Nakajima RT, Martins C. Genomic organization and comparative RNA gene in cichlid fish, with an emphasis in *Oreochromis niloticus*. Chromosome Research. 2012;**20**:279-292

[113] Turner GF. Adaptive radiation of cichlid fish. Current Biology. 2007;**17**:827-831

[114] Silva DMZA, Utsunomia R, Pansonato-Alves JC, Oliveira C, Foresti F. Chromosomal mapping of repetitive DNA sequences in five species of *Astyanax* (Characiformes, Characidae) reveals independent location of U1 and U2 snRNA sites and association of U1 snRNA and 5S rDNA. Cytogenetic and Genome Research. 2015;**146**(2):1-9. DOI: 10.1159/000438813

[115] Meyne J, Ratliff RL, Moyzis RK. Conservative of the human telomere sequences (TTAGGG)n among vertebrates. Proceedings of the National Academy of Sciences of the United States of America. 1989;**86**:7049-7053

[116] Meyne J, Baker RJ, Hobart HH, Hsu TC, Ryder OA, Ward OG, et al. Distribution of non-telomeric sites of the (TTAGGG)n telomeric sequence in vertebrate chromosomes. Chromosoma. 1990;**99**:3-10

[117] Chew JSK, Oliveira C, Wright JM, Dobson MJ. Molecular and cytogenetics analysis of the telomeric (TTAGG)n repetitive sequence in the Nile tilapia, *Oreochromis niloticus* (Teleostei: Cichlidae). Chromosoma. 2002;**111**:45-52

[118] Reed KM, Phillips RB. Molecular cytogenetic analysis of the couble-CMA3 chromosome of lake trout, *Salvelinus namaycush*. Cytogenetics and Cell Genetics. 1995;**70**:104-107

[119] Albuín M, Martinez P, Sánchez L. Localization of the telomeric sequence (TTAGGG)n in four salmonid species. Genome. 1996;**39**:1035-1038

[120] Garrido-Ramos MA, Herran R, Ruiz Rejon C, Ruiz Rejon M. A satellite DNA of Sparidae family (Pisces, Perciformes) associated with telomeric sequences. Cytogenetics and Cell Genetics. 1998;**83**:3-9

[121] Scacchetti PC, Utsunomia R, Pansonato-Alves JC, et al. Extensive spreading of interstitial telomeric sites on the chromosomes of *Characidium* (Teleostei, Characiformes). Genetica. 2015;**143**:263-270. DOI: 10.1007/s10709-014-9812-3

[122] Cioffi MB, Bertollo LAC. Initial steps in XY chromosome differentiation in *Hoplias malabaricus* and the origin of an X 1 X 2 Y sex chromosome system in this fish group. Heredity. 2010;**105**:554-561

[123] Ugarkovic D, Plohl M. Variation in satellite DNA profiles—Causes and effects. The EMBO Journal. 2002;**21**:5955-5959

[124] Datta U, Dutta P, Manda K. Cloning and characterization of a highly repetitive fish nucleotide sequence. Gene. 1998;**62**:331-336

[125] Moyer SP, Ma DP, Thomas TL, Gold JR. Characterization of a highly repeated satellite DNA from the cyprinidae fish *Notropis lutrensis*. Comparative Biochemistry and Physiology. 1988;**91B**:639-646

[126] Monaco PJ, Swan KF, Rasch EM, Musich PR. Characterization of a repetitive DNA in the unisexual fish *Poecilia formosa*. I. Isolation and cloning of the MboI family. Evolution and Ecology of Unisexual Vertebrates. 1989;**466**:123-131

[127] Grewal SIS, Jia S. Heterochromatin revised. Nature Reviews. Genetics. 2007;**8**:35-46. DOI: 10.1038/nrg2008

[128] Mestriner CA, Bertollo LAC, Galetti PM Jr. Chromosome banding and synaptonemal complexes in *Leporinus lacustris* (Pisces, Anostomidae): Analysis of a sex system. Chromosome Research. 1995;**3**:440-443

[129] Jesus CM, Galetti PM Jr, Valentini SR, Moreira-Filho O. Molecular characterization and chromosomal localization of two families of satellite DNA in *Prochilodus lineatus* (Pisces, Prochilodontidae), a species with B chromosomes. Genetica. 2003;**118**:25-32

[130] Utsunomia R, Silva DMZA, Ruiz-Ruano FJ, Araya-Jaime C, et al. Uncovering the ancestry of B chromosomes in *Moenkhausia sanctaefilomenae* (Teleostei, Characidae). PLoS One. 2016;**11**(3):e0150573. DOI: 10.1371/journal. pone.0150573

[131] Utsunomia R, Ruiz-Ruano FJ, Silva DMZA, Serrano EA, Rosa IF, et al. A glimpse into the satellite DNA library in Characidae Fish (Teleostei, Characiformes). Frontiers in Genetics. 2017;**8**:103. DOI: 10.3389/fgene.2017.00103

[132] Arai R. Fish Karyotypes: A Check List. New York: Springer Science and Business Media; 2011

[133] Nanda I, Feichtinger W, Schmid M, Schroeder JH, Zischler H, Epplen JT. Simple repetitive sequences are associated with differentiation of the sex chromosomes in the guppy fish. Journal of Molecular Evolution. 1990;**30**:456-462

[134] Devlin RH, McNeil BK, Donaldson EM. Isolation of a Y-chromosomal DNA probe capable of determining sex in Chinook Salmon. Canadian Journal of Fisheries and Aquatic Sciences. 1991;**48**:1606-1612

[135] Nakayama I, Foresti F, Tewari R, Schartl M, Chourrout D. Sex chromosome polymorphism and heterogametic males revealed by two cloned DNA probes in the ZW/ZZ fish *Leporinus elongatus*. Chromosoma. 1994;**103**:31-39

[136] Capriglione T, Morescalchi A, Olmo E, Rocco L, Stingo L, Manzo S. Satellite DNAs heterochromatin and sex chromosomes in *Chionodraco hamatus* (Channichthyidae, Perciformes). Polar Biology. 1994;**14**:285-290

[137] Stein J, Phillips RB, Devlin RH. Identification of the Y chromosome in Chinook Salmon (*Oncorhynchus tshawytscha*). Cytogenetics and Cell Genetics. 2001;**92**:108-110

[138] Haaf T, Schmid M. An early stage of ZZ/ZW sex chromosome differentiation in *Poecilia sphenops* var. *melanisticta* (Poeciliidae, Cyprinodontiformes). Chromosoma. 1984;**89**:37-41

[139] Nanda I, Schartl M, Feichtinger W, Epplen JT, Schmid M. Early stages of sex chromosome differentiation in fish as analysed by simple repetitive DNA sequences. Chromosoma. 1992;**101**:301-310

[140] Vicente VE, Jesus CM, Moreira-Filho O. Chromosomal localization of 5S and 18S rRNA genes in three *Parodon* species (Pisces, Parodontidae). Caryologia. 2001;**54**(4):365-369

[141] Pansonato-Alves JC, Serrano EA, Camacho JPM, Utsunomia R, et al. Single origin of sex chromosomes and multiple origins of B chromosomes in fish of the genus *Characidium*. PLoS One. 2014;**9**(9):e107169. DOI: 10.1371/journal.pone.0107169

[142] Camacho JP, Sharbel TF, Beukeboom LW. B-chromosome evolution. Philosophical Transactions of the Royal Society B: Biological Sciences. 2000;**355**:163-178

[143] Sampaio TR, Gouveia JG, da Silva CRM, Dias AL, da Rosa R. Molecular analysis of the B microchromosomes in *Steindacnerina insculpita* (Characiformes: Curimatidae) by microdissection. Cytogenetic and Genome Research. 2015;**146**(1):51-57. https://doi.org/10.1159/000381932

[144] Silva DMZA, Daniel SN, Camacho JPM, Utsunomia R, et al. Origin of B chromosomes in the genus *Astyanax* (Characiformes, Characidae) and the limits of chromosome painting. Molecular Genetics and Genomics. 2016:**291**(3):1407-1418. DOI: 10.1007/s00438-016-1195-y

[145] Barros AV, Sczepanski TS, Cabrero J, Camacho JP, Vicari MR, Artoni RF. Fiber FISH reveals different patterns of high-resolution physical mapping for repetitive DNA in fish. Aquaculture. 2011;**322**:47-50

[146] Oyeleye OO, Ogundeji ST, Ola SI, Omitogun OG. Basics of animal cell culture: Foundation for modern science. Biotechnology and Molecular Biology Reviews. 2016;**11**(2):6-16. DOI: 10.5897/BMBR2016.0261

[147] Paim FG. Desenvolvimento de cultura de células aderentes em peixes neotropicais e sua aplicação em estudos citogenéticos [thesis]. Botucatu, SP, Brasil: Universidade Estadual Júlio Mesquita Filho; 2018

[148] Rabova M, Monteiro R, Collares-Pereira MJ, Ráb P. Rapid fibroblast culture for teleost fish karyotyping. In: Ozouf-Costaz C, Pisano E, Foresti F, FA TL, editors. Fish Cytogenetic Techniques: Ray-Fin Fishes and Chondrichthyans. CRC Press; 2015. pp. 66-73

[149] Amemiya CT, John WB, John RG. A cell culture technique for chromosome preparation in cyprinid Fishes. Copeia. 1984;**1984**(1):232-235

[150] Zhang Q, Cooper RK, Wolters WR, Tiersch TR. Isolation, culture and characterization of a primary fibroblast cell line from channel catfish. Cytotechnology. 1998;**26**(2):83-90

[151] Wang X, Yang J, Chen X, Pan X. Establishment and characterization of a fibroblast-like cell line from *Anabarilius grahami* (Cypriniformes: Cyprinidae). Zoological Research. 2012;**33**(E5-6):E89-E97

Chapter 5

Karyology of the Bats from the Russian Far East

Uliana V. Gorobeyko and Irina V. Kartavtseva

Additional information is available at the end of the chapter

http://dx.doi.org/10.5772/intechopen.78767

Abstract

Recent studies based on morphologic and molecular genetic data have revealed quite a serious variety in the trans-Palearctic species, which brought about taxonomic status changes in 14 of 18 Russian Far Eastern bat species. Far Eastern bat status revisions resulted in species growth whose chromosome characteristics have been described either under other names or have not been studied at all. This paper has inventoried bat chromosome research in the Russian Far East and neighboring regions and has improved the accuracy of chromosome characteristics for 17 of 18 valid species today. For the first time, the karyotypes and their variation type for the valid bat species in the Russian Far East have been described.

Keywords: Chiroptera, karyotype, chromosome, nucleolar organizer regions, heterochromatic material

1. Introduction

Till the middle of the twentieth century, most of the Russian bats were considered to belong to widespread Palearctic species. Since the mid-1960s, a gradual transition from the "wide" polytypic species concept appears to be replaced by the "narrow" monotypic one [1]. This is largely due to the improved morphological data processing methods [2–4] and the use of the molecular genetic [5, 6] and the karyological [7–9] methods in bat systematics. Many of the Far Eastern bat taxa were treated formerly as eastern subspecies within polytypical trans-Palearctic species. Recently, most of the Far Eastern subspecies have been elevated to a species rank, which resulted in taxonomic status changes of 14 Far Eastern bat species [5, 6, 10–21]. However, the taxonomic status of certain forms needs to be clarified [22]. Most of these species are restricted to Northeast Asia, with the western species distribution bordering the Trans-Baikal and the Altai regions [22, 23].

Karyotype features are essential diagnostic characteristics of many mammalian species [24, 25]. Even species with similar diploid number (2n) and chromosome morphology have been shown to differ significantly in distributional patterns of nucleolar organizer regions (NOR) [26–29] and the amount and location of heterochromatic material on chromosomes [30–34].

Bats are characterized by high level of karyotype stability at the genus and low intraspecific chromosomal variability, e.g., in *Myotis* Kaup, 1929; *Eptesicus* Rafinesque, 1820; *Vespertilio* Linnaeus, 1758; *Barbastella* Gray, 1821; *Plecotus* Gray, 1866 [7, 35–39].

The so-called *Myotis-type* karyotype with 2n = 44 and fundamental number (NFa) being 50 is accepted to be the ancestral karyotype of family Vespertilionidae Gray, 1821 [37]. The chromosomal arms are usually numbered using Bickham's scheme, in which ordinal numbers have been assigned to all the autosomal arms based on GTG-banding patterns [40].

The position and number of the nucleolus organizer regions (NORs) and the amount and location of heterochromatic material (C-band) on chromosomes of many vespertilionid species have been shown to represent species-specific characteristics. The sequential staining methods (G-band; NOR; C-band) revealed karyological differences in species of the same karyotype [7, 8, 39, 41–45].

Chromosomal studies of the Far Eastern bats were initiated by N.N. Vorontsov [35] and continued by his colleagues and students [46–49]. The conventional staining of 10 bat species karyotypes was described. Differential staining (NOR and C-band) was reported for two species, *Plecotus ognevi* Kishida, 1927 and *Eptesicus nilssonii* Keyserling & Blasius, 1839.

Species composition revision of the Far Eastern bats caused an increase in the number of species, whose chromosomal characteristics were reported either under the wrong species names or were not studied at all.

The paper presents an inventory of available karyological data on bats from the Russian Far East and neighboring regions. It provides revision of specified chromosomal characteristics of 18 valid bat species from the Russian Far East. The karyotype descriptions of valid Far Eastern bat species and their chromosomal variability are given for the first time.

2. Karyotypes of Far Eastern bat

Table 1 shows valid Russian Far Eastern bat species. The columns represent species belonging to geographically various regions. The last one gives the species names describing the karyotypes. The table demonstrates the level of karyological knowledge available of certain bat species in every region studied. European and Northeastern Asian karyotype species have been studied to the fullest extent possible. Less data have been obtained regarding karyotype species in Siberia and the Russian Far East.

To illustrate the intrageneric and intraspecific variability of the Russian Far Eastern bat karyotypes based on data available, **Table 2** is drawn, which made it possible to compare chromosome characteristics of a similar Far Eastern bat species from different geographic

http://dx.doi.org/10.5772/intechopen.78767

Valid species								Formerly named in sources
Europe		Siberia		Russian Far East		Northeast Asia		
Myotis nattereri	E	*Myotis bombinus*	No	*Myotis bombinus*	No	*Myotis bombinus*	J	*Myotis nattereri*
–		*Myotis ikonnikovi*	No	*Myotis ikonnikovi*	FE	*Myotis ikonnikovi*	J	*Myotis ikonnikovi*
–		*Myotis longicaudatus*	No	*Myotis longicaudatus*	No	*Myotis longicaudatus*	J	*Myotis frater*
Myotis capaccinii	E	–		*Myotis macrodactylus*	FE	*Myotis macrodactylus*	J K	*Myotis capaccinii*
Myotis daubentonii	E	*Myotis daubentonii*	No	*Myotis petax*	FE	*Myotis petax*	K	*Myotis daubentonii*
		Myotis petax	No					
Myotis brandtii	E	*Myotis brandtii*	No	*Myotis gracilis*	No	*Myotis gracilis*	K	*Myotis brandtii*
		Myotis sibirica	S	*Myotis sibirica*	FE			
Plecotus auritus	E	*Plecotus ognevi*	S	*Plecotus ognevi*	FE	*Plecotus sacrimontis*	J	*Plecotus auritus*
		Plecotus auritus	No	*Plecotus sacrimontis*	No			
–		–		*Barbastella darjelingensis*	No	*Barbastella darjelingensis*	J	*Barbastella leucomelas*
–		–		*Pipistrellus abramus*	No	*Pipistrellus abramus*	J C K	*Pipistrellus abramus*
Vespertilio murinus	E	*Vespertilio murinus*	S	*Vespertilio murinus*	FE	*Vespertilio murinus*	no	*Vespertilio murinus*
–		*Vespertilio sinensis*	No	*Vespertilio sinensis*	FE	*Vespertilio sinensis*	J	*Vespertilio orientalis*
Hypsugo savii	E	–		*Hypsugo alashanicus*	FE	*Hypsugo alashanicus*	K	*Pipistrellus savii*
Eptesicus nilssonii	E	*Eptesicus nilssonii*	No	*Eptesicus nilssonii*	FE	*Eptesicus nilssonii*	J	*Eptesicus nilssonii*
–		–		*Murina ussuriensis*	No	*Murina ussuriensis*	J	*Murina aurata*
–		*Murina hilgendorfi*	S	*Murina hilgendorfi*	FE	*Murina hilgendorfi*	J	*Murina leucogaster*
Miniopterus schreibersii	E	–		*Miniopterus fuliginosus*	No	*Miniopterus fuliginosus*	J C T M	*Miniopterus schreibersii*

Notes: The geographical regions with the names abbreviated karyotypes investigated: E—Europe, S—Siberia, FE—Far East, J—Japan, C—China, K—Korea, T—Thailand, M—Malaysia.

Sources for species of Europe: [7, 39, 42, 44, 50], of Siberia: [47, 51], of the Far East—see **Table 2**. "no"—unknown.

Table 1. Valid species of the Far Eastern bats and their karyological studies.

Valid species	Species named in sources	Reg	2n	NFa	M-SM (large + medium + small)	ST	A	X	Y	NOR	Diff. stain.	N	Ref.
Vespertilionidae Gray 1821—common bats													
Myotis bombinus	*M. nattereri*	J	44	50	3 + 0 + 1	–	17	SM	–	–	C	1f	[41]
	M. n. bombinus	J	44	50	3 + 0 + 1	–	17	M	A	11 cmc	C, G	1m	[43]
Myotis ikonnikovi	*M. hosonoi*	J	44	52	5 + 0 + 0	–	16	SM	A	–	–	1m	[52]
	M. hosonoi	J	44	52	3 + 0 + 2	–	16	SM	A	–	–	2m 1f	[53]
	M. hosonoi	J	44	50	3 + 0 + 1	–	17	SM	A	–	C, G	10m 14f	[41]
	M. hosonoi	J	44	50	3 + 0 + 1	–	17	M-SM	–	–	C, G, Q	5m 3f	[54]
	M. ikonnikovi	FE	44	50	3 + 0 + 1	–	17	SM	–	–	–	1f	[47]
	M. hosonoi	J	44	52	3 + 0 + 2	–	16	M	A	5 cmc	G	2m 1f	[43]
Myotis longicaudatus	*M. frater kaguyae*	J	44	50	3 + 0 + 2	–	16	SM	A	–	C, G	6m	[41]
	M. frater	J	44*	50	–	–	–	M-SM	SM	–	C	–	[55]
	M. frater	J	44	50	3 + 0 + 1	–	17	M-SM	–	–	C, G, Q	3m 4f	[54]
	M. frater	J	44	52	3 + 0 + 2	–	16	M	ST	13 cmc	C, G	3m 4f	[43]
Myotis macrodactylus	*M. capaccinii*	FE	44	50	3 + 0 + 1	–	17	M	A	–	–	1m	[46]
	M. macrodactylus	J	44	52	3 + 0 + 2	–	16	SM	A	–	–	2m 2f	[53]
	M. macrodactylus	J	44 + B	52	3 + 0 + 2	–	16	SM	A	–	–	5m 5f	[56]
	M. macrodactylus	K	44	50	3 + 0 + 1	–	17	SM	A	–	–	2m 3f	[57]
	M. macrodactylus	J	44	52	3 + 0 + 2	–	16	SM	A	–	C, G	4m 6f	[41]
	M. macrodactylus	J	44	50	3 + 0 + 1	–	17	M-SM	–	–	C, G, Q	8m 2f	[55]
	M. macrodactylus	J	44	52	3 + 0 + 2	–	16	M-SM	SM	–	C	–	[54]
	M. macrodactylus	K	44	52	3 + 0 + 2	–	16	M-SM	M-SM	–	–	5m	[58]
	M. macrodactylus	J	44	52	3 + 0 + 2	–	16	M	A	6 cmc	G	7m 5f	[43]

Valid species	Species named in sources	Reg	2n	NFa	M-SM (large + medium + small)	ST	A	X	Y	NOR	Diff. stain.	N	Ref.
Myotis petax	*M. daubentonii*	FE	44*	50	3 + 0 + 1	–	17	M	A	–	–	1m 2f	[47]
	M. daubentonii	K	44	52	3 + 0 + 2	–	16	M	A	–	–	2m	[58]
Myotis sibirica	*M. brandtii*	S	44*	50	3 + 0 + 1	–	17	M	A	–	–	2m	[47]
	M. brandtii	FE	44*	50	3 + 1 + 0	–	17	M	A	–	–	1m 1f	[48]
Myotis gracilis	*Myotis mystacinus gracilis*	K	44	50	3 + 0 + 1	–	17	M-SM	A	–	–	2m	[58]
Plecotus ognevi	*P. auritus*	FE	32	50	9 + 0 + 1	–	5	SM	–	4	C	1f	[47]
	P. auritus	S	32	50	9 + 0 + 1	–	5	SM	A	–	G, Q, FISH	1m	[51]
	P. ognevi	FE	32	50	9 + 0 + 1	–	5	SM	A	–	–	1m	[49]
Plecotus sacrimontis	*P. auritus sacrimontis*	J	32	50	9 + 0 + 1	–	5	SM	A	–	–	2f	[53]
	P. a. sacrimontis	J	32*	–	–	–	–	–	–	–	–	1m 1f	[59]
	P. a. sacrimontis	J	32	50	9 + 0 + 1	–	5	M	A	4 cmc	G	1m 3f	[43]
Barbastella darjelingensis	*B. leucomelas darjelingensis*	J	32	50	10	–	5	SM	A	–	–	1m	[60]
	B. leucomelas	J	32	50	10	–	5	SM	A	–	–	–	[61]
	B. l. darjelingensis	J	32	50	9 + 0 + 1	–	5	M	A	5 cmc	G	2m 1f	[43]
Pipistrellus abramus	*P. abramus*	J	26	44	6 + 4 + 0	–	2	A	M	–	–	2m	[52]
	P. abramus	J	26	44	6 + 4 + 0	–	2	A	A	–	–	3f	[53]
	P. abramus	J	26	44	6 + 4 + 0	–	2	A	A		G	4m 3f	[62]
	P. abramus	J	26*	–	–	–	–	–	–	–	–	1m 1f	[59]
	P. abramus	J	26	44	10 + 0 + 0	–	2	A	A	–	C	–	[55]
	P. abramus	J	26	44	6 + 4 + 0	–	2	ST	–	–	C, G, Q	3m 7f	[54]
	P. abramus	K	26	44	8 + 0 + 0	2	2	A	A	–	–	1m	[58]
	P. abramus	J	26	44	10 + 0 + 0	–	2	A	A	1 int	C, G	7m 3f	[43]
	P. abramus	C	26	44	10 + 0 + 0	–	2	A	A	–	C, G	9m 6f	[63]
	P. abramus	C	26	44	10 + 0 + 0	–	2	A	A	–	–	2m 2f	[64]
	P. abramus	C	26	44	10 + 0 + 0	–	2	A	A	–	C, G	1m 7f	[65]

Valid species	Species named in sources	Reg	2n	NFa	M-SM (large + medium + small)	ST	A	X	Y	NOR	Diff. stain.	N	Ref.
Vespertilio murinus	*V. murinus*	S	38	50	6 + 0 + 1	–	11	M	A	–	–	2m	[35]
	V. murinus	E	38	50	6 + 0 + 1	–	11	M	–	–	G, Q	1m 1f	[66]
	V. murinus	E	38*	50	–	–	–	–	–	2 int	–	1m	[42]
	V. murinus	FE	38*	50	6 + 0 + 1	–	11	M	–	–	–	1f	[47]
	V. murinus	S	38	50	6 + 0 + 1	–	11	M	A	–	G, Q, FISH	1m	[51]
	V. murinus	FE	38	50	6 + 0 + 1	–	11	M	A	–	–	1m 1f	[49]
Vespertilio sinensis	*V. superans*	FE	38	50	6 + 0 + 1	–	11	M	A	–	–	3m 2f	[35]
	V. orientalis	J	38	50	6 + 0 + 1	–	11	SM	A	–	–	–	[61]
	V. orientalis	J	38	50	6 + 0 + 1	–	11	SM	A	–	C	3m 7f	[67]
	V. superans	J	38	50	6 + 0 + 1	–	11	M-SM	A	–	C	–	[55]
	V. superans	J	38	54	6 + 0 + 3	–	9	SM	Dot	–	C, G	5m 5f	[68]
	V. orientalis											3m 5f	
	V. superans	FE	38*	50	6 + 0 + 1	–	11	M	A	–	–	2m 2f	[47]
	V. superans	J	38	50	6 + 0 + 1	–	11	M	A	2 int	G	3m 5f	[43]
	V. superans	J	38	50	6 + 0 + 1	–	11	M	A	–	C, T, Q, FISH	1m	[69]
Hypsugo alashanicus	*P. savii koreensis*	K	44	50	3 + 0 + 1	–	17	M	–	–	–	2f	[57]
	P. savii	FE	44*	50	3 + 0 + 1	–	17	M	–	–	–	1f	[47]
	P. koreensis	K	44	50	3 + 0 + 1	–	17	M-SM	A	–	–	3m	[58]

Valid species	Species named in sources	Reg	2n	NFa	M-SM (large + medium + small)	ST	A	X	Y	NOR	Diff. stain.	N	Ref.
Eptesicus nilssonii	*E. parvus*	J	50	48	–	–	–	–	–	–	–	1f	[59]
	E. nilssonii	E	50*	48	–	–	24	–	–	–	–	–	[70]
	E. nilssonii	FE	50	48	–	–	24	M	–	1 int	C	2f	[47]
	E. nilssonii	J	50	48	–	–	–	–	–	–	–	–	[71]
	E. n. parvus	J	50	50	–	1	23	M-SM	A	–	T, Q, FISH	2m 1f	[69]
	E. nilssonii	FE	50*	48	–	–	24	M	A	–	–	1m 1f	[48]
	E. nilssonii	E	50	48	–	–	24	M-SM	–	1 int	G	1f	[44]
Murina hilgendorfi	*M. leucogaster hilgendorfi*	J	44	50	3 + 0 + 1	–	17	M	A	–	–	1m	[53]
	M. leucogaster	J	44	58	3 + 0 + 1	4	13	SM	A	–	–	–	[60]
	M. l. hilgendorfi	J	44	56	3 + 0 + 1	3	14	SM	A	–	C, G	2m	[72]
	M. leucogaster	FE	44	50	2 + 1 + 1	–	17	SM	A	–	–	1m	[47]
	M. hilgendorfi	S	44	56	3 + 0 + 1	3	14	SM	A	–	G, Q, FISH	1m	[51]
Murina ussuriensis	*M. aurata*	J	44	60	3 + 0 + 2	4	12	SM	A	–	–	–	[61]
	Murinus auratus ussuriensis	J	44	50	3 + 0 + 1	–	17	M	A	–	–	1m	[59]
	M. aurata ussuriensis	J	44	56	3 + 0 + 1	3	14	SM	A	–	C, G	1m 1f	[72]
	M. sylvatica	J	44	56	3 + 0 + 1	3	14	–	–	num. cmc	–	1m 2f	[43]

Valid species	Species named in sources	Reg	2n	NFa	M-SM (large + medium + small)	ST	A	X	Y	NOR	Diff. stain.	N	Ref.
Miniopteridae Dobson 1835—Bent-winged Bats													
Miniopterus fuliginosus	*M. schreibersii fuliginosus*	J	46	52	2 + 1 + 1	–	18	SM	A	–	–	3m 1f	[73]
	M. s. fuliginosus	J	46	52	2 + 1 + 1	–	18	SM	A	–	–	8m 6f	[53]
	M. schreibersii	M	46	50	2 + 0 + 1	–	19	SM	A	–	–	1m 1f	[74]
	M. s. haradai	T	46	52	2 + 1 + 0	1	18	SM	A	–	–	2m	[70]
	M. s. fuliginosus	J	46	50	2 + 0 + 1	–	19	M	A	1cmc 1int	G	1m 1f	[43]
	M. schreibersii	T	46	50	2 + 0 + 1	–	19	SM	A	–	–	1f	[75]
	M. fuliginosus	C	46	50	2 + 0 + 1	–	19	SM	A	–	G, FISH	–	[76]
	M. fuliginosus	C	46	50	2 + 1 + 0	–	19	SM	–	–	C, G	1f	[77]
	M. schreibersii	C	46	50	2 + 1 + 0	–	19	SM	A	–	–	1m	[65]

*The chromosome image is not shown at the sources; "–", no data.

Columns: reg.—geographical regions, M-SM—number of biarmed chromosome pairs (size: large + medium + small); Diff. stain.—differential staining of chromosome (G, C, etc.); NOR—AgNOR-banding (cmc—centromere-cap NORs, int—interstitial NORs); N—number of specimens examined (f—female, m—male); Ref.—literature sources.

Morphology of chromosomes: M—metacentric, SM—submetacentric, M-SM—biarmed, ST—subtelocentric, A—acrocentric, dot—dot-like chromosome.

Geographical regions abbreviations: E—Europe, S—Siberia, FE—Far East, J—Japan, C—China, K—Korea, T—Thailand, M—Malaysia.

Table 2. Far Eastern bats karyological data.

regions for the first time and to reveal availability or lack of this variability. For simplicity sake, three size groups have been introduced to analyze size variability of two-arm (M-SM) chromosomes: large, medium-sized, and small ones, with their respective karyotype numbers assigned. This allowed us to show the karyotype variability based on this feature. Besides, **Table 2** also shows the previous study of the species by using different sequential staining methods for the chromosomes, thus making it possible to differentiate species with a similar chromosome formula.

Integrated data on the karyotypes, extent of their studies, and chromosome variability of the Russian Far Eastern bats are provided below.

2.1. Family Vespertilionidae Gray, 1821: common bats

2.1.1. Genus Myotis Kaup, 1829: mouse-eared bats

All *Myotis* species have similar karyotypes: 2n = 44 [7, 35, 39, 42, 46]. The fundamental arms number varied from 50 to 52 in different studies. This is due to the fact that some authors accounted for short euchromatic arms on the seven autosomal pairs [7, 39], while the others described this one as an acrocentric [41, 43, 46–48, 54–57]. For some authors, NFa also covered the additional heterochromatic short arms on 24 or 25 pairs of acrocentrics [41, 43, 52, 53, 55, 57]. The species of genus *Myotis* showed the centromere-cap NORs (cmcNORs), with the distributional pattern of NORs in *Myotis* karyotype being species-specific [7, 39, 42].

The amount and location of C-band in Eurasian *Myotis* chromosomes varies intra- and inter-specifically [39, 41, 43, 54, 55]. Eurasian *Myotis* species proved to have small heterochromatic segments close to the centromere on most of the chromosomal arms. Certain *Myotis* species show a distinct intercalary heterochromatic segments found in the proximal part of chromosome 15, in the vicinity of the centromere on chromosomal arm 16, and in the short arm of the X-chromosome adjacent to the centromere [39]. The size and morphology of Y-chromosome were species-specific and depended on amount of heterochromatic material in chromosome [39]. Asian bat species karyotypes have a distinctly pronounced totally heterochromatic short arm on one of the dot-like chromosomes 24 and 25. There might be a tiny second arm in several species or a large heterochromatic secondary arm of the same size as the euchromatic arm [39, 41, 43].

The genus *Myotis* is the most frequently found bats genus in the Russian Far East, with seven recorded species. Of these, six species are also spread in Northeast Asia and five species are common in Siberia. Karyotype of one species was reported found in Siberia. The karyotypes of four *Myotis* species studied are common for the Russian Far East. The karyotypes of five *Myotis* species were described from Northeast Asia.

M. bombinus **Thomas, 1906.** The karyotypes were described from Japan species. The cmc-NORs were shown to be located in 11 autosomal pairs: from 7 to 15, 19, and 22. The hetero-chromatic short arms on chromosome 25 of *M. bombinus* were tiny or absent at all [41].

M. ikonnikovi **Ognev, 1912.** The karyotypes were reported from Japan and the Russian Far East. It was shown that the cmcNORs were located in 7, 13, 14, 22, and 23 autosomal pairs.

Intraspecific variability is likely to exist here regarding the large heterochromatic short arms on the 25 autosomal pairs [41].

***M. longicaudatus* Ognev, 1927.** The karyotype was studied using the Japan species. The cmcNORs were located on 13 autosomal pairs: from 8 to 11, from 13 to 15, and from 18 to 23. The morphology of Y-chromosome seems to vary from acrocentric [41] and subtelocentric [43] to submetacentric [55]. The morphology of chromosome 25 appears to vary from acrocentric to submetacentric due to the presence or absence of heterochromatic short arms [41, 43].

***M. macrodactylus* (Temminck, 1840).** The karyotype was described using Northeast Asia and the Russian Far East specimens (**Figure 1**). The cmcNORs were located on 18–23 autosomal pairs. The morphology of chromosome 25 seems to vary from acrocentric chromosome in *M. macrodactylus* from the Russian Far East [46], Korea [58], and Japan [54] to metacentric chromosome in other Japanese *M. macrodactylus* [41, 43, 53, 55, 56]. The presence of one B-chromosome for *M. macrodactylus* from Japan has been showed [56].

***M. petax* Hollister, 1912.** The conventionally stained karyotype of *M. petax* was studied from Korea and the Russian Far East. The Korean and Far Eastern *M. petax* appeared to differ by a number of small biarmed chromosomal pairs.

***M. sibirica* Kaschenko, 1905.** The routinely staining karyotype was described from Siberia and the Russian Far East. No pronounced differences in the karyotypes of Siberian and Far Eastern *M. sibirica* have been found.

***M. gracilis* Ognev, 1927.** The conventionally stained karyotype of *M. gracilis* was studied from Korea.

So, out of seven Far Eastern species, *Myotis* karyotype has been studied for all of them. Although all *Myotis* species have similar karyotypes with 2n = 44, the distributional pattern of NORs and the amount and location of heterochromatic material in the karyotype are the

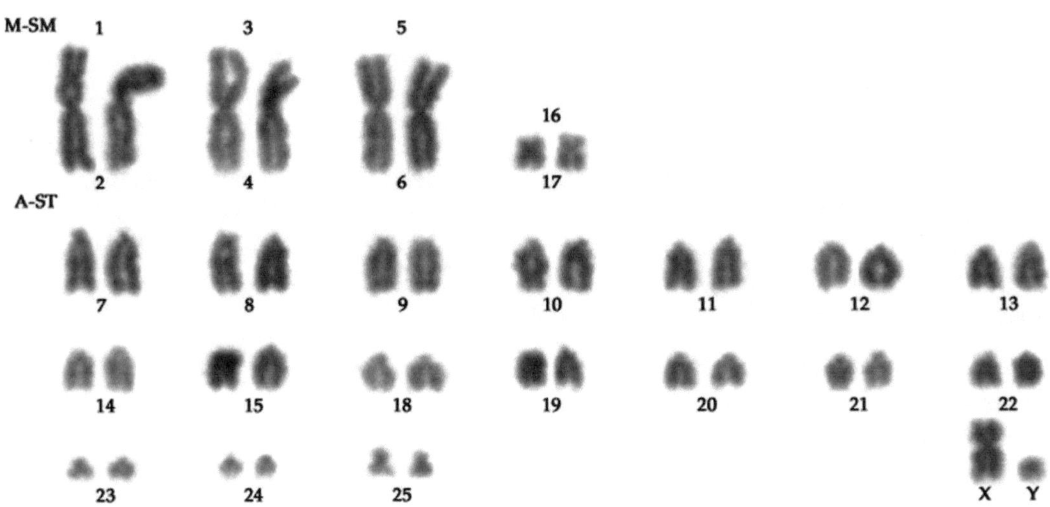

Figure 1. Karyotype of *Myotis macrodactylus* from the Russian Far East [our data].

most important differentiating characteristics for the *Myotis* species. Various levels of the data studied for differently staining *Myotis* chromosomes from various Northeastern regions make it impossible to do species comparative analysis based on the above features.

2.1.2. Genus Plecotus Gray, 1866: Old World long-eared bats

The species of genus *Plecotus* are characterized by a karyotype with 2n = 32, NFa = 50 [43, 47, 49, 66]. The distributional pattern of NORs is a centromere-cap NOR (cmcNORs) [42, 43, 47].

There are two species of *Plecotus* in the Russian Far East: *P. ognevi* and *P. sacrimontis*.

P. ognevi **Kishida, 1927.** The karyotype of *P. ognevi* was described from the Russian Far East (**Figure 2**). Four NORs were found belong to acrocentric chromosomes of *P. ognevi*; but it was impossible to determine the numbering of these chromosomal arms according to *Myotis-type* karyotype because of G-banding failure [47]. The distributional patterns of heterochromatic material in karyotype were shown: large heterochromatic segments were found in all biarmed autosomal pairs, while small C-band emerged in the most acrocentric chromosomes except the first pair [47].

G-staining, Q-banding, and Zoo-FISH of Siberian *P. ognevi* karyotypes were studied. A pericentric inversion or centromere shift on the smallest metacentric *P. ognevi* chromosome 16/17 using the HSA 16 probe was revealed, which accounted for the differences between G-banding patterns and the homologous *Myotis* species chromosome [51].

P. sacrimontis **G. Allen, 1908**. Karyotype of *P. sacrimontis* was reported from Northeast Asia. NORs were located on chromosomes 20, 22, 23, and 24 [43], while the European species *P. auritus* Linnaeus, 1758 showed NORs on 20, 22, 24, and 25 autosomal pairs [42].

So, all *Plecotus* species have similar karyotypes with 2n = 32, NFa = 50. *P. auritus* and *P. sacrimontis* had different NORs distribution on chromosomes. For *P. ognevi*, it was impossible to determine the numbering and NOR location on chromosomal arms. Heterochromatic distribution pattern in karyotype was studied only for *P. ognevi* from the Russian Far East, thus making it impossible to compare data from various species and regions.

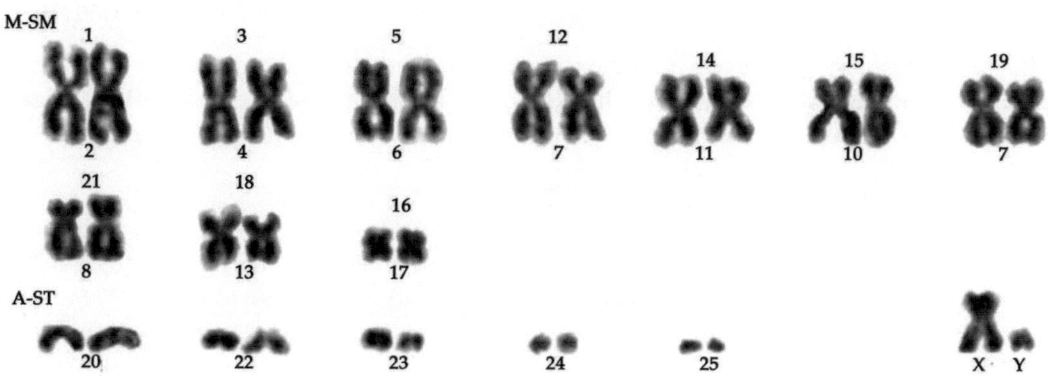

Figure 2. Karyotype of *Plecotus ognevi* from the Russian Far East. The figure was previously published in our paper, see [49].

2.1.3. Genus Barbastella Gray, 1821: barbastelles

Karyotype of *Barbastella* is similar to that of the *Plecotus* karyotype: 2n = 32, NFa = 50. The distributional pattern of NORs is cmcNORs [43].

There is only one species of genus *Barbastella* in the Russian Far East—***B. darjelingensis* Hodgson, 1855.** It can be found exclusively on the island of Kunashir [23, 78]. The chromosomal set was reported only from *B. darjelingensis* from Northeast Asia. Five NORs were found on 21–25 autosomal pairs of standard *Myotis-type* karyotype [43].

2.1.4. Genus Pipistrellus Kaup, 1829: pipistrelles

The genus *Pipistrellus* is characterized by considerable variability of 2n and NFa [35].

There is one pipistrelles species inhabiting the Russian Far East, i.e., ***P. abramus* Temminck, 1840.** Karyotype of *P. abramus* was described from Northeast Asia. Unlike other pipistrelles, *P. abramus* has low number 2n and NFa (2n = 26, NFa = 44) due to centric fusions. Chromosome rearrangements complexity makes it impossible to identify the chromosomal arms by G-banding that were involved in composition of 5 out of 10 biarmed pairs of *P. abramus* karyotype. Therefore, the numbering of *P. abramus* chromosomes differs from *Myotis-type* karyotype [43, 54, 63, 65].

The distributional pattern of NORs is interstitial (intNORs). The large NOR was located in secondary constriction (SC) of five metacentric pairs consisting of 14 and 7 autosomal pairs of *Myotis-type* karyotype [43].

The intraspecific variations of sex chromosomes in karyotype of especially *P. abramus* were likely to be found. Many researchers identified X chromosome morphology as a medium-sized acrocentric, while the X chromosome of the *P. abramus* from Fukuoka prefecture (Japan) was described as subtelocentric [54]. The Y chromosome of *P. abramus* was usually characterized as the smallest acrocentric, while the Y chromosome of the same species from Gunma prefecture (Japan) was described as a small metacentric [52].

High intraspecific variability of heterochromatic material seems to be specific of the *P. abramus* karyotype. This variability for *P. abramus* from Northeast Asia is presented in **Table 3**.

The *P. abramus* karyotype is described only from Northeastern Asia specimens, which can be possibly accounted for by existing intraspecific variability based on morphology of sex chromosomes, number and localization of structural heterochromatin in karyotype.

2.1.5. Genus Vespertilio Linnaeus, 1758: particolored bats

All specimens of genus *Vespertilio* showed the karyotypes with 2n = 38, NFa = 50 [35, 44, 79]. All *Vespertilio* species showed location of two large intNORs in the SC of 15 and 23 autosomal pairs [42, 43]. There are two *Vespertilio* species in the Russian Far East—*V. murinus* and *V. sinensis*.

***V. murinus* Linnaeus, 1758** is the trans-Palearctic bat species, whose karyotype was described from Europe, Siberia, and the Russian Far East. The NOR distributional pattern was reported from Europe [42]. The chromosome characteristics show stability across the entire area of its distribution (**Figure 3**).

***V. sinensis* Peters, 1880** belongs to the East Asian bat species. The karyotype was described from Northeast Asia and the Russian Far East. NFa = 54 was shown to characterize some specimens from Japan, probably due to the fact that certain researchers included small heterochromatic secondary arms on the two smallest acrocentric in NFa [68]. The distributional pattern of NORs was reported from Northeast Asia [43]. The significant intraspecific polymorphism seems to exist in regard to amount and location of heterochromatic material in karyotype of Japanese *V. sinensis* (**Table 4**).

2n	NFa	No. chromosomal arms														Reg.	Ref.
		1	2	3	4	5	6	7	8	9	10	11	12	*X*	*Y*		
26	44	+	+	+	+	+	+	+	o	o	o	+	o	+	•	J	[55]
26	44	+	+	+	+	+	o	o	o	o	+	+	o	+	–	J	[54]
26	44	+	+	+	+	o	o	o	o	o	o	o	o	o	–	J	[43]
26	44	+	+	+	+	+	o	+	o	+	+	+	+	+	•	C	[63]
26	44	+	+	+	+	o	o	o	o	o	+	+	o	+	–	C	[65]

Note: o—totally euchromatic chromosomes; +—heterochromatic band in vicinity of the centromere; •—totally heterochromatic chromosomes.
Geographical regions abbreviations: J—Japan, C—China.

Table 3. Intraspecific variations of heterochromatic material in karyotypes of ***Pipistrellus abramus***.

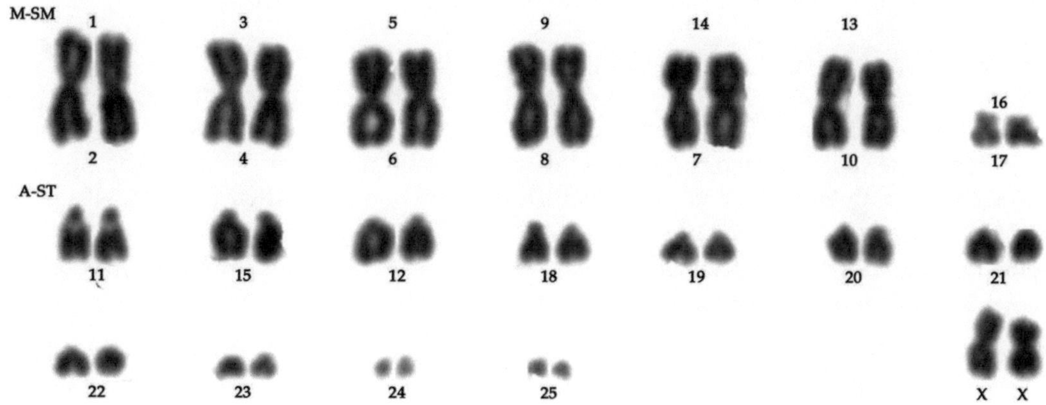

Figure 3. Karyotype of *Vespertilio murinus* from the Russian Far East. The figure previously was published in our paper, see [49].

2n	NFa	No. chromosomal arms																				Ref.
		1/2	3/4	5/6	13/7	11/8	9/ 10	16/ 17	12	14	*15	18	19	20	21	22	*23	24	25	X	Y	
38	50	o	o	o	o	o	o	o	o	o	int	+	+	+	+	+	+	•	•	+	•	[67]
38	50	o	o	+	o	+	o	+	+	+	+	o	+	+	+	+	+	•	•	+	•	[55]
38	54	+ int	+	+ int	+	+	+	+	+	+	+ int	+	+	+	+	+	+	•	•	+	•	[68]
38	50	+	+	+ int	+ int	+	+	+	+	o	+ int	+	+	+	+	+	+	•	•	+	•	[69]

Note: o—totally euchromatic chromosomes; +—heterochromatic band in vicinity of the centromere; •—totally heterochromatic chromosomes; *—secondary construction on the chromosome.

Table 4. Intraspecific variations of heterochromatic material in karyotypes of Japanese ***Vespertilio sinensis***.

The localization of telomeric sequences $(TTAGGG)_n$ was described by FISH for *V. sinensis* from Japan. Hybridization signals were observed at both ends of all *V. sinensis* chromosomes along with very faint and small-sized interstitial signals that were also present at centromeric sites of all seven biarmed chromosomes. Large and intense hybridization signals revealed themselves at the centromeric regions in eight pairs of acrocentric autosomes (18–25) and the Y chromosome of *V. sinensis*. It is interesting to note that C-band of the smallest acrocentric pair 25 and of the Y chromosome displayed a complete hybridization, while interstitial C-band in 5/6, 7/13, and 15 autosomal pairs in *V. sinensis* exhibited no hybridization. Internal telomeric sequences were observed in the heterochromatic regions or satellite DNA on chromosomes that may indicate recent chromosomal rearrangements occurred in the evolution process [69].

While the chromosome characteristics of *V. murinus* show stability across the entire area of its distribution, the karyotype of *V. sinensis* seems to have a significant intraspecific polymorphism regarding the content of structural heterochromatin in the karyotype.

2.1.6. Genus Hypsugo Kolenati, 1856: high pipistrelles

The diploid number and fundamental number of genus *Hypsugo* chromosomes noticeably vary due to the centric fusions as well as inversions and centromere shift [44]. The *Hypsugo* species show both intNORs and cmcNORs. The *H. savii* Bonaparte, 1837 (2n = 44, NFa = 50) and *H. eisentrauti* (Hill, 1968) (2n = 42, NFa = 58) exhibit only one intNORs in SC of chromosome 15, while *H. crassulus* Thomas, 1904 (2n = 30, NFa = 56) possesses cmcNORs on chromosomes 3 and 19 and in proximal part of chromosome 15/25 [44].

There is only one *Hypsugo* species found in the Russian Far East—*H. alashanicus* Bobrinskoy, 1926. This karyotype was described from Northeast Asia and the Russian Far East 2n = 44, NFa = 50.

2.1.7. Genus Eptesicus Rafinesque, 1820: serotines

Karyotypes of all autosomes belonging to *Eptesicus* species can be characterized as acrocentric: 2*n* = 50, NFa = 48 [8, 36, 44].

There is only one *Eptesicus* species found in the Russian Far East—*E. nilssonii* Keyserling & Blasius, 1839. *E. nilssonii* species distribution is trans-Palearctic. The karyotype of *E. nilssonii* was reported from Europe, Northeast Asia, and the Russian Far East (**Figure 4**). 2n and NFa are the same for most of the studied *E. nilssonii* excepting this one from Hokkaido with one biarmed autosomal pair in karyotype [69]. The large intNORs is located on secondary constriction in chromosome 15 [44, 47].

The amount and location of heterochromatic material in karyotype was described for *E. nilssonii* from the Russian Far East. There were small C-bands on all chromosomes pairs, and the fourth largest pair showed a large interstitial heterochromatic segment. The SC on chromosome 15 showed C-band [47].

The chromosome characteristics of *E. nilssonii* including distributional pattern of NORs show stability across the entire area of its distribution. Structural heterochromatin distribution pattern was studied only for the Far Eastern *E. nilssonii,* which prevented us from evaluating variability of this feature.

2.1.8. *Genus Murina* Gray, *1842: tube-nosed bats*

The karyotypes of tube-nosed bats do not differ from 2n = 44 [72, 80, 81], while NFa varies from 50 to 60 probably due to subtelocentric pairs produced by the pericentric inversions [7, 71, 79]. The distributional pattern of NORs is cmcNORs [43, 80]. There are two *Murina* species in the Russian Far East, which are *M. hilgendorfi* and *M. ussuriensis.*

***M. ussuriensis* Ognev, 1914**. Karyotype of *M. ussuriensis* was described from Japan. With the known localization type, the localization of multiple cmcNORs on chromosomes has not been determined yet because G-banding has not been done [43].

The amount and location of heterochromatic material in *M. ussuriensis* karyotype were described from Japan. The autosomal pairs 5/6, 16/17, 20, 24 and X chromosome showed small centromeric C-bands, while the Y chromosome was totally heterochromatic. The interstitial faintly stained C-band was revealed in the distal part of X chromosome [72].

***M. hilgendorfi* Gray, 1842.** Karyotype of *M. hilgendorfi* was described from Siberia, Northeast Asia, and the Russian Far East region (**Table 2**).

Karyotype of one specimen from Primorsky Velican cave (the Russian Far East) was clearly different from other *M. hilgendorfi* ones by the number of large biarmed pairs: there were only two large metacentric pairs, one medium-sized submetacentric pair being approximately equal to a long arm of large metacentric pair and one small metacentric pair [47]. The same karyotype was previously described for a tube-nosed bat from Thailand [70]. It was originally reported as *M. leucogaster* Milne-Edwards, 1872, though later the bat was redefined as *M. harrisoni* Csorba & Bates, 2005 [82]. However, karyotypes of other specimens of *M. harrisoni* [81, 83] and *M. leucogaster* [84] exhibited karyotype with three large biarmed chromosomal pairs.

The amount and location of heterochromatic material in karyotype were shown for *M. hilgendorfi* from Japan. There were small C-band close to centromere on chromosomes 5/6, 16/17, 20, 24 and X chromosome with totally heterochromatic Y chromosome [72].

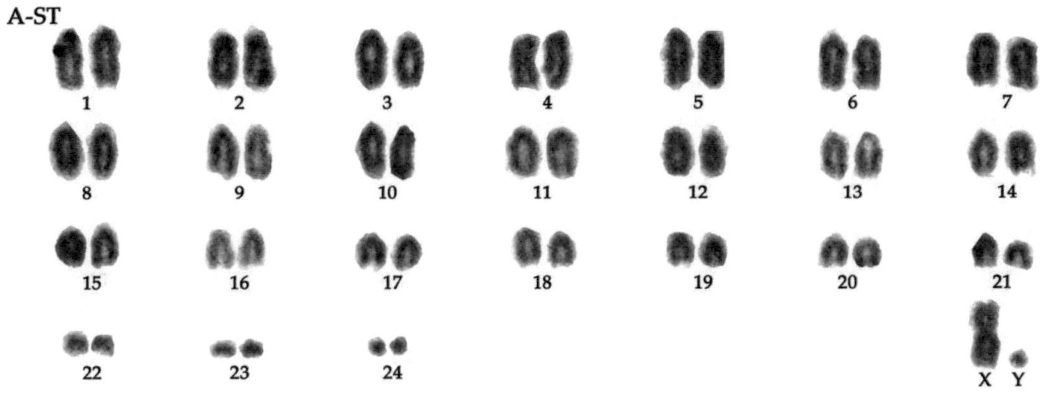

Figure 4. Karyotype of *Eptesicus nilssonii* from the Russian Far East. The figure previously was published in paper [48].

The location of structural heterochromatin of *M. ussuriensis* and *M. hilgendorfi* from Japan scarcely differs from each other. *M. hilgendorfi* karyotype with two large metacentric pairs, one medium-sized submetacentric pair and 1 small metacentric pair described from the Russian Far East, seemed to be either in error or an isolated case that requires verification.

2.2. Family Miniopteridae Dobson, 1875: bent-winged bats

2.2.1. Genus Miniopterus Bonaparte, 1837: bent-winged bats

Karyotypes of bent-winged bats are clearly different from standard *Myotis-type* karyotype due to chromosomal rearrangements. By using GTG-staining and FISH methods, the biarmed chromosome 3/4 of *Myotis-type* karyotype was shown to be similar to two acrocentric pairs of *Miniopterus,* due to centric fissions the metacentric pair 16/17 assumed the shape of an acrocentric, and the acrocentric pair 12 became biarmed due to pericentric inversions, with the G-banding pattern of 7 and 10 autosomal arms being different from standard *Myotis-type* karyotype [76].

There is one species of the monotypic family Miniopteridae found in the Russian Far East that is *M. fuliginosus* Hodgson, 1835. Karyotype (2n = 46, NFa = 50–52) was described from Northeast Asia.

The *M. fuliginosus* seems to exhibit intraspecific polymorphism by the number of biarmed autosomal pairs. Karyotype with two large and one small biarmed pairs is most common. *M. fuliginosus*, with its mostly encountered karyotype, was found in Malaysia, Thailand, China, and Japan [43, 74–76]. Karyotype with two large and one medium biarmed chromosomal pair was described from China [65, 77]. Karyotype of *M. fuliginosus* from Thailand was similar to the previous one with one exception: it had one subtelocentric pair [71]. Karyotype with two large, one medium, and one small biarmed pairs was described from Japan [53, 73].

One cmcNORs was shown to be located on 20 autosomal pair and one intNOR is located on chromosome 23 in the *M. fuliginosus karyotype* from Japan [43]. The small C-band close to centromere was described to be located on all chromosomal pairs of Chinese *M. fuliginosus* [77].

So, *M. fuliginosus* from Northeastern Asia seems to be characterized by intraspecific chromosome polymorphism based on the number of autosomal pairs.

3. Conclusion

For the first time, the references' analysis undertaken enabled us to demonstrate the extent of chromosome characteristics studied for bats from the Russian Far East. It also illustrated the nature of the intrageneric and intraspecific chromosome variability of the bats from the Russian Far East.

The data available enable us to suggest *Miniopterus fuliginosus, Murina hilgendorfi,* and some *Myotis* species to show intraspecies chromosome polymorphism regarding biarmed autosomal pairs. Intraspecies variability could be fairly assumed to exist as regards X,Y chromosomes in *P. abramus, M. longicaudatus* and *M. macrodactylus* karyotypes from Northeastern Asia. A significant intraspecies polymorphism regarding structural heterochromatin in a karyotype seems to be available in *V. sinensis, P. abramus,* and *Myotis* species. Such important characteristic as the amount and localization of cmcNORs on chromosomes has been very irregularly studied for the Far Eastern bat species, which restricts our ability to compare data from different regions. There is not enough data to compare *Barbastella* and *Hypsugo* species in terms of their karyotype chromosome characteristics.

Thus, one might make a conclusion that karyotypes of the majority bats from the Russian Far East and Siberia still remain to be studied. The bats from Northeastern Asia and Europe have their bats' chromosome characteristics somewhat more fully explored, though we still have considerable gaps in our knowledge of karyotypes for certain bats' species.

Acknowledgements

The reported study was funded by the Russian Foundation for Basic Research according to the research project № 18-34-00285.

Author details

Uliana V. Gorobeyko* and Irina V. Kartavtseva

*Address all correspondence to: ekz.bio@ya.ru

Federal Scientific Center of the East Asia Terrestrial Biodiversity Far Eastern Branch of Russian Academy of Sciences, Vladivostok, Russian Federation

References

[1] Strelkov PP. The crisis of the polytypic species concept as illustrated by the genus *Plecotus*. Plecotus et al. 2006;**9**:3-7

[2] Tiunov MP. Bats of the Russian Far East. Vladivostok: Dal'nauka Press; 1997. 134 pp

[3] Kruskop SV. Towards the taxonomy of the Russian *Murina*. Russian Journal of Theriology. 2005;**4**(2):135-140. DOI: 10.15298/rusjtheriol.4.2.01

[4] Bulkina TM, Kruskop SV. Search for morphological differences between genetically distinct brown long-eared bats (*Plecotus auritus* s. lato, Vespertilionidae). Plecotus et al. 2009;**11-12**:3-13

[5] Matveev VA, Kruskop SV, Kramerov DA. Revalidation of *Myotis petax* Hollister, 1912 and its new status in connection with *M. daubentonii* (Kuhl, 1817) (Vespertilionidae, Chiroptera). Acta Chiropterologica. 2005;**7**(1):23-37. DOI: 10.3161/1733-5329(2005)7[23:ROMPHA]2.0.CO;2

[6] Spitzenberger F, Strelkov PP, Winkler H, Haring E. A preliminary revision of the genus *Plecotus* (Chiroptera, Vespertilionidae) based on genetic and morphological results. Zoologica Scripta. 2006;**35**(3):187-230. DOI: 10.1111/j.1463-6409.2006.00224.x

[7] Volleth M, Heller KG. Phylogenetic relationships of vespertilionid genera (Mammalia: Chiroptera) as revealed by karyological analysis. Zeitschrift für Zoologische Systematik und Evolutionsforschung. 1994;**32**:11-34. DOI: 10.1111/j.1439-0469.1994.tb00467.x

[8] Kearney TC, Volleth M, Contrafatto G, Taylor PG. Systematic implications of chromosome GTG-band and bacula morphology for southern African *Eptesicus* and *Pipistrellus* and several other species of Vespertilioninae (Chiroptera: Vespertilionidae). Acta Chiropterologica. 2002;**4**(1):55-76. DOI: 10.3161/001.004.0107

[9] Volleth M, Son NT, Wu Y, Li Y, Yu W, Lin LK, Arai S, Trifonov V, Liehr T, Harada M. Comparative chromosomal studies in *Rhinolophus formosae* and *R. luctus* from China and Vietnam: Elevation of *R. l. lanosus* to species rank. Acta Chiropterologica. 2017;**19**(1):41-50. DOI: 10.3161/15081109ACC2017.19.1.003

[10] Maeda K. Review on the classification of little tube-nosed bats, *Murina aurata,* group. Mammalia. 1980;**44**(4):531-551. DOI: 10.1515/mamm.1980.44.4.531

[11] Horaček I, Hanak V. Comments on the systematics and phylogeny of *Myotis nattereri* (Kuhl, 1818). Myotis. 1984;**21-22**:20-29

[12] Yoshiyuki M. A Systematic Study of the Japanese Chiroptera. Tokyo: National Science Museum; 1989. p. 242

[13] Horaček I. Status of *Vesperus sinensis* Peters, 1880 and remarks on the genus Vespertilio. Vespertilio. 1997;**2**:59-72

[14] Horaček I, Hanak V, Gaisler J. 2000. Bats of the Palearctic : A taxonomic and biogeographic review. In: Proceedings of the 8th European Bat Research Symposium (EBRS'00); January 2000; Krakow. Krakow: Institute of Systematics and Evolution of Animals PAS; 2000. pp. 11-157. DOI: 10.13140/2.1.4099.2643

[15] Kawai K, Nikaido M, Harada M, Matsumura S, Lin LK, Wu Y, Hasegawa M, Okada N. The status of the Japanese and east Asian bats of the genus *Myotis* (Vespertilionidae) based on mitochondrial sequences. Molecular Phylogenetics and Evolution. 2003;**28**(2):297-307. DOI: 10.1016/S1055-7903(03)00121-0

[16] Tian L, Liang B, Maeda K, Metzner W, Zhang S. Molecular studies on the classification of *Miniopterus schreibersii* (Chiroptera: Vespertilionidae) inferred from mitochondrial cytochrome b sequences. Folia Zoologica. 2004;**3**(53):303-311

[17] Kawai K, Kondo N, Sasaki N, Fukui D, Dewa H, Satô M, Yamaga Y. Distinguishing between cryptic species *Myotis ikonnikovi* and *M. brandtii gracilis* in Hokkaido, Japan: Evaluation of a novel diagnostic morphological feature using molecular methods. Acta Chiropterologica. 2006;**8**(1):95-102. DOI: 10.3161/1733-5329(2006)8[95:DBCSMI]2.0.CO;2

[18] Benda P, Dietz C, Andreas M, Hotovy J, Lucan RK, Maltby A, Meakin K, Truscott J, Vallo P. Bats (Mammalia: Chiroptera) of the Eastern Mediterranean and Middle East. Part 6. Bats of Sinai (Egypt) with some taxonomic, ecological and echolocation data on that fauna. Acta Societatis Zoologicae Bohemicae. 2008;**72**:3-103

[19] Artyushin IV, Bannikova AA, Lebedev VS, Kruskop SV. Mitochondrial DNA relationships among North Palaearctic *Eptesicus* (Vespertilionidae, Chiroptera) and past hybridization between Common Serotine and Northern Bat. Zootaxa. 2009;**2262**:40-52. DOI: 10.11646/zootaxa.2262.1.2

[20] Kruskop SV, Borisenko AV, Ivanova NV, Lim BK, Eger JL. Genetic diversity of northeastern Palaearctic bats as revealed by DNA barcodes. Acta Chiropterologica. 2012;**14**(1):1-14. DOI: 10.3161/150811012X654222

[21] Ruedi M, Csorba G, Lin LK, Chou CH. Molecular phylogeny and morphological revision of *Myotis* bats (Chiroptera: Vespertilionidae) from Taiwan and adjacent China. Zootaxa. 2015;**3920**(1):301-342

[22] Kruskop SV. Order Chiroptera. In: Pavlinov IY, Lissovsky AA, editors. The Mammals of Russia: A Taxonomic and Geographic Reference. Moscow: KMK Sci. Press; 2012. 604 p

[23] Tiunov MP. Distribution of the bats in Russian Far East (Problems and questions). In: Proceedings of the Japan-Russia Cooperation Symposium on the Conservation of the Ecosystem; 2011; Okhotsk. Sapporo. 2011. pp. 359-369

[24] Vorontsov NN. The importance of chromosomal sets for mammalian taxonomy. Bulletin of the Moscow Society of Naturalists. 1958;**6**(2):5-36

[25] Matthey R. The chromosome formulae of eutherian mammals. In: Cytotaxonomy and Vertebrate Evolution. London: Academic Press; 1973. pp. 531-616

[26] Korablev VP. Localization of nucleolar organizer regions in mammals. In: Questions of Evolutionary Zoology and Mammalian Genetics. Vladivostok. 1987. pp. 37-44

[27] Sánchez A, Burgos M, Jiménez R, Díaz de la Guardia R. Variable conservation of nucleolus organizer regions during karyotypic evolution in Microtidae. Genome. 1990;**33**(1): 119-122

[28] Boeskorov GG, Kartavtseva IV, Zagorodnyuk IV, Belyanin AN, Lyapunova EA. Nucleolus organizer regions and B-chromosomes of wood mice (Mammalia, Rodentia, Apodemus). Russian Journal of Genetics. 1995;**31**(2):185-192

[29] Kartavtseva IV. Karyosystematics of Wood and Field Mice (Rodentia: Muridae). Vladivostok: Dal'nauka Press; 2002. 144 p

[30] Hsu TC, Arrighi FE. Distribution of constitutive heterochromatin in mammalian chromosomes. Chromosoma. 1971;**34**(3):243-253. DOI: 10.1007/BF00286150

[31] White MJD. Animal Cytology and Evolution. 3rd ed. Cambridge: Cambridge University Press; 1973. 961 p

[32] Prokofyeva-Belgovskaya AA. Heterochromatic regions of chromosomes: Structure and functions. Biology Bulletin Reviews. 1977;**38**(5):735-757

[33] Prokofyeva-Belgovskaya AA. Heterochromatic Regions of Chromosomes. Moscow: Nauka; 1986. 431 p

[34] Korobitsyna KV, Korablev VP. The intraspecific autosome polymorphism of *Meriones tristrami* Thomas, 1892 (Gerbillinae, Cricetidae, Rodentia). Genetica. 1980;**52-53**(1):209-221. DOI: 10.1007/BF00121829

[35] Vorontsov NN, Radjabli SI, Volobuev VT. The comparative karyology of the vespertilionid bats, Vespertilionidae (Chiroptera). In: Vorontsov NN, editor. The Mammals (Evolution, Karyology, Taxonomy, Fauna). Novosibirsk: Nauka Press; 1969. pp. 16-21

[36] Baker RJ. Karyotypic trends in bats. In: Biology of Bats. Vol. 1. New York: Academic Press; 1970. pp. 65-95

[37] Baker RJ, Bickham JW. Karyotypic evolution in bats: Evidence of extensive and conservative chromosomal evolution in closely related taxa. Systematic Zoology. 1980;**29**(3):239-253. DOI: 10.1093/sysbio/29.3.239

[38] Baker RJ, Qumsiyeh MB, Hood CS. Role of chromosomal banding patterns in understanding mammalian evolution. In: Genoways HH, Current Mammalogy. Boston: Springer; 1987. pp. 67-96. DOI: 10.1007/978-1-4757-9909-5_2

[39] Volleth M, Heller KG. Variations on a theme: Karyotype comparison in Eurasian *Myotis* species and implications for phylogeny. Vespertilio. 2012;**16**:329-350

[40] Bickham JW. Banded karyotypes of 11 species of American bats (genus *Myotis*). Cytologia. 1979;**44**:789-797. DOI: 10.1508/cytologia.44.789

[41] Harada M, Yoshida TH. Karyological study of four Japanese *Myotis* bats (Chiroptera, Mammalia). Chromosoma (Berlin). 1978;**65**:283-291. DOI: 10.1007/BF00327623

[42] Volleth M. Differences in the location of nucleolus organizer regions in European vespertilionid bats. Cytogenetics and Cell Genetics. 1987;**44**:186-197. DOI: 10.1159/000132371

[43] Ono T, Obara Y. Karyotypes and Ag-NOR variations in Japanese vespertilionid bats (Mammalia: Chiroptera). Zoological Science. 1994;**11**(3):473-484

[44] Volleth M, Bronner G, Gopfert MC, Heller KG, von Helversen O, Yong HS. Karyotype comparison and phylogenetic relationships of *Pipistrellus*-like bats (Vespertilionidae; Chiroptera; Mammalia). Chromosome Research. 2001;**9**:25-46. DOI: 10.1023/A:1026787515840

[45] Volleth M, Heller KG, Fahr J. Phylogenetic relationships of three "Nycticeiini" genera (Vespertilionidae, Chiroptera, Mammalia) as revealed by karyological analysis. Mammalian Biology—Zeitschrift für Säugetierkunde. 2006;**71**(1):1-12. DOI: 10.1016/j.mambio.2005.09.001

[46] Volobuev VT, Strelkov PP. The karyotypes identity in the genus *Myotis*. Russian Journal of Zoology. 1971;**4**(12):1892-1894

[47] Korablev BP, Yakimenko LV, Tiunov MP. Karyotypes of bats in the Russian far east. In: Kryukov AP, Chelomina GN, Pavlenko MV, editors. The Present-Day Approached to Studies of Variability: Collection of Scientific Papers. Vol. 1989. Vladivostok: The Far Eastern Branch Academy of Sciences of the USSR; 1989. pp. 95-98

[48] Kartavtseva IV, Dokuchayev NE. Studying chromosomes of two types of bats in Kamchatka. In: Proceedings of the Biological Diversity of Siberian Animals; 28-30 October 1998; Tomsk. Tomsk: Del'taplan; 1998. pp. 67-68

[49] Kartavtseva IV, Gorobeiko UV, Tiunov MP. The current status of chromosomal investigations of bats (Chiroptera) from the Russian far east. Russian Journal of Zoology. 2014;**93**(7):887-900. DOI: 10.7868/S0044513414070083

[50] Arslan A, Zima J. Karyotypes of the mammals of Turkey and neighbouring regions: A review. Folia Zoologica -Praha-. 2014;**63**(1):1-62. DOI: 10.25225/fozo.v63.i1.a1.2014

[51] Kulemzina AI, Nie W, Trifonov VA, Staroselec Y, Vasenkov DA, Volleth M, Yang F, Graphodatsky AS. Comparative chromosome painting of four Siberian Vespertilionidae species with Aselliscus stoliczkanus and human probes. Cytogenetic and Genome Research. 2011;**134**:200-205. DOI: 10.1159/000328834

[52] Tsuchiya K, Harada M, Yosida TH. Karyotypes of four species of bats collected in Japan. Annual Report of National Institute of Genetics (Japan). 1972;**2**:50-51

[53] Harada M. Chromosomes of nine chiropteran species in Japan. La Kromosomo. 1973;**91**:2885-2895

[54] Ando K, Harada M, Uchida TA. A karyological study on five Japanese species of *Myotis* and *Pipistrellus*, with special attention to composition of their C-band materials. Journal of the Mammalogical Society of Japan. 1987;**12**(1-2):25-29

[55] Ando K, Tagawa T, Uchida TA. The C-banding pattern of 6 Japanese species of vespertilionine bats (Mammalia: Chiroptera). Experientia. 1980;**36**:653-653. DOI: 10.1007/BF01970118

[56] Obara Y, Tomiyasu T, Saitoh K. Chromosome studies in the Japanese vespertilionid bats: I. Karyotypic variation in *Myotis macrodactylus* Temminck. Japanese Journal of Genetics. 1976;**51**(3):201-206. DOI: 10.1266/jjg.51.201

[57] Park SR, Won PO. Chromosomes of Korean bats. Journal of the Mammalogical Society of Japan. 1978;**7**:199-203. DOI: 10.11238/jmammsocjapan1952.7.199

[58] Yoo DH, Yoon MH. A karyotypic study on six Korean vespertilionid bats. Korean. Journal of Zoology. 1992;**35**(4):489-496

[59] Tsuchiya K. A contribution to the chromosome study in Japanese mammals. Proceedings of the Japan Academy. 1979;**55B**(4):191-195. DOI: 10.2183/pjab.55.191

[60] Uchida TA, Ando K. Karyotype analysis in Chiroptera (I): Karyotype of the eastern barbastelle, *Barbastella leucomelas darjelingensis* and comments on its phylogenetic position. Science Bulletin of the Faculty of Agriculture. Kyushu University. 1972;**26**(1/4):393-398. DOI: 10.15017/23098

[61] Ando K, Tagawa T, Uchida TA. Considerations of karyotypic evolution within Vespertilionidae. Experientia. 1977;**33**:877-879. DOI: 10.1007/BF01951257

[62] Obara Y, Tomiyasu T, Saitoh K. Chromosome studies in the Japanese vespertilionid bats: G-banding pattern of *Pipistrellus abramus* Temminck. Proceedings of the Japan Academy. 1976;**52**(7):383-386

[63] Lin LK, Motokawa M, Harada M. Karyological study of the house bat *Pipistrellus abramus* (Mammalia: Chiroptera) from Taiwan with comments on its taxonomic status. The Raffles Bulletin of Zoology. 2002;**50**(2):507-510

[64] Wu Y, Harada M, Li Y. Karyology of seven species bats from Sichuan, China. Acta Theriologica Sinica. 2004;**4**(1):30-35

[65] Wu Y, Motokawa M, Li YC, Harada M, Chen Z, Lin LK. Karyology of eight species of bats (Mammalia: Chiroptera) from Hainan Island, China. International Journal of Biological Sciences. 2009;**5**:659-666. DOI: 10.7150/ijbs.5.659

[66] Volleth M. Chromosomal homologies of the genera *Vespertilio*, *Plecotus* and *Barbastella* (Chiroptera: Vespertilionidae). Genetica. 1985;**66**:231-236. DOI: 10.1007/BF00128044

[67] Obara Y, Saitoh K. Chromosome studies in the Japanese vespertilionid bats: IV. Karyotypes and C-banding pattern of *Vespertilio orientalis*. The. Japanese Journal of Genetics. 1977;**52**(2):159-161. DOI: 10.1266/jjg.52.159

[68] Harada M, Ando K, Uchida TA, Takada S. Karyotypic evolution of two Japanese *Vespertilio* species and its taxonomic implication (Chiroptera: Mammalia). Caryologia. 1987;**40**(3):175-184. DOI: 10.1080/00087114.1987.10797821

[69] Ono T, Yoshida MC. Differences in the chromosomal distribution of telomeric $(TTAGGG)_n$ sequences in two species of the vespertilionid bats. Chromosome Research. 1997;**5**:203-212. DOI: 10.1023/A:1018403215999

[70] McBee K, Bickham JW, Yehbutra S, Nabhitabhata J, Schlitter DA. Standard karyology of nine species of vespertilionid bats (Chiroptera: Vespertilionidae) from Thailand. Annals of Carnegie Museum. 1986;**55**(5):95-116

[71] Ono T, Yoshida MC. Banded karyotype of *Eptesicus nilssonii parvus* (Mammalia: Chiroptera). Chromosome Information Service. 1995;**59**:9-21

[72] Harada M, Ando K, Uchida TA, Takada S. A karyological study on two Japanese species of *Murina* (Mammalia: Chiroptera). Journal of the Mammalogical Society of Japan. 1987;**1-2**:15-23. DOI: 10.11238/jmammsocjapan1987.12.15

[73] Tsuchiya K. Chromosomes of two insectivorous bat species from Japan. Journal of the Mammalogical Society of Japan. 1971;**5**(3):114-116

[74] Harada M, Kobayashi T. Studies on the small mammal fauna of Sabah, East Malaysia. II. Karyological analysis of some Sabahan mammals. Contributions from the Biological Laboratory. 1980;**26**:83-95

[75] Lin LK, Motokawa M, Harada M. Karyology of ten vespertilionid bats (Chiroptera: Vespertilionidae) from Taiwan. Zoological Studies. 2002;**41**(4):347-354

[76] Ao L, Gu X, Feng Q, Wang J, O'Brien PC, Fu B, Mao X, Su W, Wang Y, Volleth M, Yang F, Nie W. Karyotype relationships of six bat species (Chiroptera, Vespertilionidae) from China revealed by chromosome painting and G banding comparison. Cytogenetic and Genome Research. 2006;**115**(2):145-153. DOI: 10.1159/000095235

[77] Li N, Ao L, He SY, Gu XM. G-bands and C-bands in 3 species of Vespertilionidae. Chinese Journal of Zoology. 2007;**42**(2):96-101

[78] Selezneva TA, Tiunov MP. *Barbastella leucomelas* (Cretzschmar, 1826)—A new species for the fauna of the Russian Far East. In: Proceedings of VIII Meeting of Theriological Society; 31 January—2 February 2007; Moscow. Moscow: KMK Scientific Press Ltd; 2007. p. 443

[79] Bickham J. Chromosomal variation and evolutionary relationships of vespertilionid bats. Journal of Mammalogy. 1979;**60**(2):350-363. DOI: 10.2307/1379807

[80] Volleth M. Karyotype analysis of *Murina suilla* and *Phoniscus atrox* from Malaysia (Chiroptera: Murininae, Kerivoulinae). Lynx (Praha). 2006;**37**:275-284

[81] Son NT, Csorba G, Tu VT, Thong DV, Wu Y, Harada M, Oshida T, Endo H, Motokawa M. A new species of the genus *Murina* (Chiroptera: Vespertilionidae) from the Central Highlands of Vietnam with a review of the subfamily Murininae in Vietnam. Acta Chiropterologica. 2015;**17**(2):201-232. DOI: 10.3161/15081109ACC2015.17.2.001

[82] Francis CM, Eger JL. A review of tube-nosed bats (*Murina*) from Laos with description of two new species. Acta Chiropterologica. 2012;**14**(1):15-38. DOI: 10.3161/150811012X654231

[83] Wu Y, Motokawa M, Li YC, Harada M, Chen Z, Yu WH. Karyotype of Harrison's tube-nosed bat *Murina harrisoni* (Chiroptera: Vespertilionidae: Murininae) based on the second specimen recorded from Hainan Island, China. Mammal Study. 2010;**35**(4):277-279. DOI: 10.3106/041.035.0407

[84] Gu XM. The karyotypes of six species of bats from Guizhou. Chinese Journal of Zoology. 2006;**41**(5):112-116

Chapter 6

Resolving Paradoxes of Robertsonian Translocations

Natalia V. Kovaleva

Additional information is available at the end of the chapter

http://dx.doi.org/10.5772/intechopen.79237

Abstract

Since Robertsonian translocations (ROB) are essential in the etiology of congenital malformations and reproductive disorders, it is natural to assume that they represent a thoroughly studied subject. However, on closer inspection, there are poorly studied areas within this field. The aim of this report is to present results of a comprehensive analysis of available data collected by researchers worldwide that allows a new look at the problems mentioned above. There were determined rates and spectrums of ROB in the general population and in patients with reproductive disorders. The comprehension of a female-based sex ratio (male-to-female ratio) among newborn carriers of balanced nonhomologous ROB in the general population leads to a conclusion on the mechanism of sex-specific correction of translocation trisomy, which might explain both inexplicably low occurrence of rob-associated uniparental disomy and phenomenon of "non-Mendelian-inheritance." The data obtained indicate that female ROB carriers are at a much higher risk of uniparental disomy compared to male ROB carriers. In the majority of asymptomatic male carriers of homologous translocation/isochromosome (HT), spermatogenesis is not impaired. An analysis of sex ratio among ill-defined HT carriers showed a difference between patients with Prader-Willi syndrome and Angelman syndrome, indicating different mechanisms of HT formation.

Keywords: Robertsonian translocations, isochromosomes, sex ratio, uniparental disomy, non-Mendelian inheritance, reproductive disorders, Prader-Willi syndrome, Angelman syndrome

IntechOpen

1. Introduction

Robertsonian translocations (ROBs) are common structural chromosome rearrangement in humans. Since they are central in the etiology of congenital malformations and reproductive disorders, it is natural to assume that they represent a thoroughly studied subject. However, on closer inspection, there are poorly studied areas within this field. Surprisingly, exact rates of ROB carriers were determined neither among consecutive newborns nor among patients with reproductive disorders. The literature reiterates the information on tenfold, or even more than tenfold, increase in the rate of ROB carriers among patients with reproductive disorders compared to the general population. In addition, the quoted rates among newborns vary depending on the source that the authors cite [1–3]. Another omission in the area under consideration is the lack of systematic comparative analysis of the ROB spectrum in various carrier groups. The phenomenon of exceptional rarity of some nonhomologous rearrangements was not given due attention. There are some enigmatic problems in the field not yet resolved. One of them, unusual segregation of maternally transmitted translocations, has been discussed for the last five decades [4–6]. Another, established more recently, is the unexpectedly low incidence of ROB-associated uniparental disomy among carriers of balanced rearrangement [7]. The epidemiology of Robertsonian homologous translocations (HTs)/isochromosomes, due to their rarity, has largely not been investigated. The aim of this report is to present results of a comprehensive analysis of available data collected by researchers worldwide that allows a new look at the problems mentioned above.

2. Materials and methods

Study groups: newborns, prenatal diagnoses for indications other than familial rearrangement (the main indication for prenatal testing was advanced maternal age, and the transmitting parent was defined following detection of a rearrangement in the fetus), spontaneous abortuses with regular and translocation trisomy for chromosome 13 and chromosome 14, carriers of *rob* (13;14)-associated maternal uniparental disomy for chromosome 14, couples with reproductive disorders, patients with male infertility, and ill-defined carriers of homologous translocation/isochromosome (listed in Additional files S1–S8: Tables S1–S10; Additional file 11: Supplemental References, available either on request or from https://www.researchgate.net/profile/Natalia_Kovaleva/contributions).Methods:meta-analysis of data retrieved from published studies. Only reports on ROB carriers of known sex were selected for the study. The data were analyzed using two packages of statistical programs: one of which utilized procedures of traditional approach and the other one utilized procedures of a modern Bayes approach. Guided by modern recommendations for the statistical analysis, we did not limit ourselves to the null hypothesis significance testing based on the p-values but also calculated the 95% confidence intervals (CIs) for proportions and their ratios. StatXact, the world's most expansive toolkit for exact nonparametric inference StatXact-8 (Cytel Co., USA), was used. To construct CIs for the proportion ratios, the method of variance estimates recovery (MOVER) algorithm implemented in the program MOVER-R.xls (http://medicine.cf.ac.uk/primary-care-public-health/resources/) was used.

3. Results and discussion

3.1. Determination of exact rates and spectrums of ROB in the general population and in patients with reproductive disorders

The rates, spectrum, and parental origin of major nonmosaic balanced rearrangements in the general population are presented in the Additional files, Tables S1–S4. Statistical analysis showed distributions of nonhomologous ROBs from all studied groups to be homogenous in all combinations; therefore, both control groups were aggregated for further analysis. In the aggregated control (**Table 1**), the results seem to be in accordance with current views on the spectrum of individual ROBs, with the overwhelming majority of rob(13;14) 71%, followed by rob(14;21) 12%; the remaining translocations are rare or exceptionally rare; rob(15;21) and rob (13;21) were detected once each (0.4%). The total frequency of all translocations, calculated for newborns, is 1.06‰ with 95% CI from 0.8 to 1.3‰.

Data on patients with reproductive disorders are presented in Additional files S1–S3: Tables S1–S3. The distribution of translocations in couples with reproductive disorders (**Table 2**) is generally similar to that observed in the aggregated control group. However, the proportion of rob(13;14) is much less in couples with habitual abortion (139/245 = 57%, with 95% CI of 51–63%), while the proportion of homologous translocations is high (24/245 = 10%, with CI of 7–14%). The overall rate of ROB carriers among couples with infertility is 3.6‰ (95% CI of 2.8–4.1‰), and 4.8‰ (95% CI of 4.2–5.5‰) among couples with multiple miscarriages. These values, as can be seen, do not exceed ten times the value in general population. A high incidence of ROB was found among patients with male infertility, 7.1‰ (95% CI of 6.2–8.2‰). Among couples with miscarriages, there is a difference between males and females by proportions of carriers of rob(14;15) (1 and 6%, correspondingly) and carriers of rob(14;21) (5 and 14%, correspondingly). There is a difference between couples with habitual abortion and couples with infertility in involving of chromosome 22 into nonhomologous rearrangements (32/245 = 14% with 95% CI of 9–18% vs. 4/110 = 4.2% with 95% CI of 1.5–9%), as well as with patients with male infertility (2/201 = 1.3% with CI 0.3–3.5%). In addition, among HT patients with habitual miscarriages, most are carriers of translocations/isochromosomes 22 (7 of 24).

Of note is the extremely low frequency of *rob*(13;21); no carriers of this translocation were found in the newborn population, while among patients with habitual miscarriage, with a fourfold concentration of translocation carriers, only one carrier of *rob*(13;21) was found. This suggests one possible mechanism, a negative selection against certain types of translocations.

This hypothesis is consistent with the data of British authors [9] who reported the discovery of three constitutional *rob*(15;21) carriers among 95 children with acute lymphoblastic leukemia. It was proposed that the mechanism of triggering the neoplastic process is chromotrypsis. The authors concluded that in carriers of this rearrangement, the risk of the disease is 2700 times higher than in the general population. Interestingly, their assumption of a population frequency of *rob*(15;21) of about 1 per 100,000 newborns is very close to the real value presented in this paper.

Indeed, *rob*(15;21) appeared to be a very rare rearrangement, which is clearly not supported by natural selection: in the normal population, only one carrier of a *rob*(15;21) was detected (sex

Studied group	Gender	Number of tested patients	Number of ROB carriers	Nonhomologous rearrangements										Homologous rearrangements				
				13;14	13;15	13;21	13;22	14;15	14;21	14;22	15;21	15;22	21;22	13;13	14;14	15;15	21;21	22;22
Newborns (Table S1)	♂♂	33,371	24 (25)[a]	18	0	0	0	2	1	1	0	1	1	0	0	0	0	0
	♀♀	31,534	38 (39)[b]	33	0	0	1	0	4	0	0	0	0	0	0	0	0	0
	ns	28,811	34[c]	26	0	0	0	0	6	0	1	1	0	0	0	0	0	0
	Total	93,716	96 (98)[a,b]	77	0	0	1	2	11	1	1	2	1	0	0	0	0	0
Prenatal diagnoses (Table S3)	♂♂		56	35	4	0	1	0	12	3	0	1	0	0	0	0	0	0
	♀♀		86	55	5	1	3	4	6	5	0	2	3	1	0	1	0	0
	Total		142 (143)[c]	90	10[c]	1	4	3	18	8	0	3	3	1	0	1	0	0
Total			238 (241)	164	9	1	5	5	28	8	1	5	4	1	0	1	0	0

[a]Including carrier of 45,XY,tdic(D;D).
[b]Including carrier of 45,XX,t(D;D).
[c]In a part of this study (Nielsen, Wohlert, 1991), gender was reported (Nielsen, Sillesen, 1975); see Additional file 11: Supplemental references.

Table 1. Spectrum of Robertsonian translocations in consecutive newborns and in prenatal diagnoses for indications other than familial translocation (updated from [8]).

Patients		Number of tested patients	Number of detected carriers	Nonhomologous rearrangements										Homologous rearrangements				
				13;14	13;15	13;21	13;22	14;15	14;21	14;22	15;21	15;22	21;22	13;13	14;14	15;15	21;21	22;22
Couples with infertility (Table S5)	♂♂	15,432	91	68	5	0	0	5	11	1	0	0	1	0	0	0	0	0
	♀♀	15,468	20	12	2	0	1	1	2	0	0	1	0	1	0	0	0	0
	Total	30,900	111	80	6	0	1	6	13	1	0	1	1	1	0	0	0	0
Couples with habitual abortion (Table S6)	♂♂	25,577	86 (87)[a]	56	3	0	2[c]	1	4	4	1	5	1	2	1	2	1	3
	♀♀	25,676	159 (160)[b]	83	2	1	4[d]	9[d]	22	6	7	5	5	5	4	1	1	4
	Total	51,253	245 (248)[e]		139	1	6	10	26	10	8	10	6	7	5	3	2	7
Patients with male infertility (Table S7)	♂♂	28,112	201	140[f]	11	1	0	9	27	1	5	0	1	2[g]	2	1	0	1

[a]Including 45,XY,t(D;G) carrier.
[b]Including 45,XX,t(D;D) carrier.
[c]Including carrier of 45,XY,t(13;22), inv.(6) (Valkova, 1986).
[d]Including a carrier of 44,XX,t(13;22),t(14;15) (Sugiura-Ogasawara et al., 2008).
[e]Including carrier of t(13;14) of unknown gender.
[f]Including two patients with 45,XY,inv.(5) (Dul et al., 2012; Tuerlings et al., 1998).
[g]Carrier of 45,XY,der(13;13)/46,XY,der(13;13),der(13;13) (Tuerlings et al., 1998); see Additional file S11: Supplemental references.

Table 2. Spectrum of Robertsonian translocations in patients with reproductive disorders (updated from [8]).

not specified), while among about a twofold smaller group of patients with habitual miscarriage, eight carriers of this translocation were diagnosed. Five carriers of *rob*(15;21) were identified among patients with male factor of infertility. These observations are of significance for medical genetic counseling of the carriers. Firstly, it is necessary to find out whether the risk of leukemia varies among the carriers depending on whether this translocation is inherited or occurred de novo. Currently, such data are not available.

Based on this data review, it is evident that it is necessary to continue accumulating survey data of couples with reproductive disorders to establish the existence or absence of differences in the range of ROB both between the patient groups and the population.

3.2. The phenomenon of female predominance among carriers of ROB in the general population has promoted comprehension of both low incidence of ROB-associated uniparental disomy and transmission ratio distortion in offspring of female ROB carriers

3.2.1. The parental origin of ROB and the sex ratio among carriers in the general population and in prenatal diagnosis

The sex ratios (SR) and parental origin of major nonmosaic balanced rearrangements in the general population are presented in the Additional files, Tables S2 and S4. The observed sex ratio was 1.06 (95% CI 1.04–1.07) which correlates with population ratios worldwide (Table S2).

The majority of both RECs and ROBs detected among consecutive newborns (but not inversions) occurred de novo. Interestingly, the proportions of mutant REC and mutant ROB in newborns were similar (9/50 = 18% and 7/52 = 13%, correspondingly), despite different parental origins: RECs arise predominantly in spermatogenesis [10, 11], while ROBs arise predominantly in oogenesis [12, 13].

Some female prevalence among transmitting parents was in concordance with reported data on REC carriers (23mat/18pat), but not on carriers of ROB (24mat/21pat), since according to common conception, a twofold female predominance should be expected in this group due to reduced male fertility of ROB heterozygotes [14].

However, the most intriguing finding is the SR variability in newborns depending on the type of rearrangement (**Table 3**); there were equal numbers of REC carriers of both sexes (31 M/31F; for rates of 0.93 and 0.98‰, correspondingly) and a notable female predominance among carriers of ROB (27 M/41F, for rates of 0.77 and 1.24‰, correspondingly). The difference between the SR among carriers of ROB (0.61 with 95% CI of 0.27–1.00) and the SR among tested newborns (1.06 with CI of 1.04-1.07) was statistically significant (Bayes approach).

Analysis of the SR according to the parental origin of rearrangements showed female preponderance among ROB carriers in either maternal or paternal origin or de novo origin: 11 M/13F, 7 M/14F, and 2 M/5F, correspondingly. Among carriers identified prenatally for indications other than familial rearrangement, female-based SR was found for both maternally and paternally transmitted rearrangements: 26 M/43F and 23 M/35F, correspondingly.

Collectively, among carriers of ROB with known parental origin, there were 67 males and 105 females (SR = 0.64), a difference from the expected ratio of 1:1 was determined to be significant

Studied group	Reciprocal translocations				Robertsonian translocations				Inversions			
	Maternal origin		Paternal origin		Maternal origin		Paternal origin		Maternal origin		Paternal origin	
	♂♂	♀♀	♂♂	♀♀	♂♂	♀♀	♂♂	♀♀	♂♂	♀♀	♂♂	♀♀
Newborns (Table S4)	15	8	8	9	11	13	7	14	2	6	0	3
	23 M/17F, SR = 1.35				18 M/27F, SR = 0.67				2 M/9F			
Prenatal diagnoses (Table S5)	51	43	52	36	26	43	23	35	45	49	54	47
	103 M/79F, SR = 1.3				49 M/78F, SR = 0.63				99 M/96F, SR = 0.96			
Total	126 M/96F, SR = 1.31				67 M/104F, SR = 0.64[a]				101 M/105F, SR = 0.96			
Sex ratio with 95% CI	$_{0.92}1.22_{1.62}$				$_{0.50}0.68_{0.93}$[b]				$_{0.77}1.03_{1.39}$			

[a]Difference with the expected ratio of 1:1 is statistically significant at $p = 0.0033$ (binomial test).
[b]Difference with the expected population ratio of 1.06 is statistically significant (Bayes approach).

Table 3. Sex ratio among carriers of balanced rearrangements according to parental origin (updated from [19]).

statistically by both traditional statistics (p = 0.0033, binomial test) and by a Bayes approach (**Table 3**). Among offspring of REC carriers and carriers of inversion, SR was not different statistically from the expected ratio of 1:1. (126 M/96F, SR = 1.31 and 102 M/105F, SR = 0.96, correspondingly).

Among ROBs identified in newborns, the vast majority of the cases constitute translocations between chromosomes 13 and 14 (50 of 61). It is these rearrangements that determine unusual SR among ROB carriers: out of 50 carriers of der(13;14), 18 were males and 32 were females (SR = 0.56). A similar ratio was observed among fetuses with der(13;14): 32 male carriers and 53 female carriers (SR = 0.60). In total, SR among carriers of der(13;14) was 0.59 (50 M/85F), which is statistically significant from the expected 1:1 ratio both when using standard statistics (p = 0.001) and when using Bayes approach.

Thus, there is currently unexplained mechanism for maintaining female-biased sex ratio in carriers of ROB. A biased SR among offspring of male ROB carriers would have been explained by some meiotic process providing preferable production of X-bearing gametes with ROB. However, for female carriers, such a mechanism cannot be considered, since women produce X-bearing gametes only, and the offspring's gender is determined by male gametes. For an explanation of the discussed phenomenon, the author suggests application of the concept of sex-specific correction of initial trisomy mostly in female embryos [15, 16]. In relation to ROBs, that means the loss of the odd chromosome is not involved to the translocation. If it is true, among carriers of balanced rearrangements, female-biased SR is expected, along with male preponderance among carriers of unbalanced translocations.

3.2.2. Sex ratio among abortuses with unbalanced translocation 13 and among abortuses with unbalanced translocation 14

Carriers of an unbalanced 46,+13,der(13;14) rearrangement are rarely found among liveborns. In the population of 64,905 newborns, translocation T13 was detected in four instances; among them only 1 was identified as der(13;14). Similarly, they are rarely found at amniocentesis in the second trimester: 2 instances only among 52,965 and 31,194 tested fetuses [17, 18]. Carriers of the other unbalanced derivative of rob(13;14), i.e., translocation trisomy for chromosome 14, 46,+14,der(13;14), are unlikely to survive to a long gestation age. Therefore, aiming to obtain data on SR among carriers of T13 and/or T14, the author analyzed studies on chromosomal constitution in spontaneous abortions.

Table 4 summarizes the data from 26 surveys that detected cases of regular and/or translocation trisomy (T) of either chromosome 13 or 14 (see Additional file: Table S8). Analysis showed that among abortuses with regular T13, there were some predominance of male carriers, 75 M/63F (SR = 1.2), not statistically different from the population ratio of 1.06. In contrast, an unusual increase in the proportion of male carriers was observed among carriers of translocation T13 (17 M/3F) which might be interpreted as evidence supporting female-specific correction of translocation trisomy. Increased SR among carriers of translocation T14 in comparison with carriers of regular T14 was observed as well, with 15 M/9F (SR = 1.7) vs. 25 M/39F (SR = 0.6), correspondingly. It is quite possible that elimination of male embryos trisomic for chromosome 14 occurred at earlier stages of embryo development.

3.2.3. Sex ratio among carriers of balanced translocation 45,der(13;14), upd(14) resulted from correction of initial translocation trisomy 14

To evaluate whether a correction of translocation T14 occurs predominantly in female carriers, one may study the SR among individuals with uniparental disomy 14, upd(14). Unlike upd (13), upd(14) carriers demonstrate clinical manifestations depending on the sex of the transmitting parent and have therefore undergone cytogenetic and molecular testing. Analysis of published cases with reported sex of the carriers of upd(14) showed that of 16 patients with 45, der(13;14),upd(14), 12 were females, including 8 carriers of upd(14)mat [20–27] and 4 carriers of upd(14)pat [28–31]; the remaining 4 male patients had upd(14)mat [32–35].

It was logical to assume that in this group, incomplete correction of initial translocation trisomy 14 may take place as the result of postzygotic events, i.e., mosaicism can be found. Moreover, carriers of mosaicism were expected to be females. Accordingly, mosaicism 45,XX, der(13;14)/46,XX,der(13;14),+14 was detected in two female patients [20–21].

References[a]	**Regular trisomy**				**Translocation trisomy**			
	Chromosome 13		**Chromosome 14**		**46,+13,der(13;14)**		**46,+14,der(13;14)**	
	♂♂	♀♀	♂♂	♀♀	♂♂	♀♀	♂♂	♀♀
Additional file: Table S8	73	63	27	39	17	3	15	9
Sex ratio with 95% CIs	$_{0.8}\,1.2\,_{1.6}$		$_{0.43}\,0.7\,_{1.13}$		$_{1.8}\,4.8^{b}\,_{17.4}$		$_{0.7}\,1.7\,_{3.7}$	

[a]Only studies where trisomy for either chromosome 13 or chromosome 14 were detected.
[b]Different statistically from the expected ratio of 1.06, P (Bayes approach).

Table 4. Sex ratio in spontaneous abortions with nonmosaic regular and translocation trisomy 13 or 14 (updated from [19]).

Among carriers of other translocations with upd(14)mat, there was also a female predominance, with four females out of five patients [25, 36–39]. This observation supports the suggestion that the trisomy correction phenomenon might not be restricted to unbalanced translocation (13;14). The data obtained is of clinical significance, indicating that female ROB carriers are at a much higher risk of uniparental disomy than male ROB carriers.

3.2.4. Preferential loss of a maternal extra chromosome in female embryos as a correction mechanism leading to biparental disomy

The data obtained, while presenting evidence for sex-specific correction of trisomy as a reason for female predominance among carriers of balanced ROB, are in apparent contradiction with the data on low incidence of uniparental disomy carriers among both prenatally tested fetuses and abortuses with familial translocations. According to collective data, the incidence of translocation trisomy correction causing uniparental disomy does not exceed 1% [7]. It is understandable that so rare an event cannot cause the observed bias in the sex ratio. In turn, the low incidence of uniparental disomy due to trisomy correction is in contradiction with the data on a very high incidence of self-correction found in preimplantation embryos [40, 41].

An assumption of a special correction mechanism leading to biparental disomy might explain this contradiction. Such a mechanism, a preferential loss of maternal chromosome (and, hence, reconstitution of biparental disomy) in female embryos, was suggested as an explanation of the twofold male predominance among patients with Prader-Willi syndrome due to maternal uniparental disomy [15] (for details, see Section 4.3.2).

Preferential loss of maternal extra chromosome in carriers of inherited unbalanced translocation may be explained "topographically": in the human zygote, maternal and paternal pronuclei are separated, and this condition is preserved during some mitotic divisions. In the case of translocation trisomy (which mostly have maternal origin), a competition for spindle attachment occurs. The vast majority of human ROBs are dicentric [12]. The dicentric structure allows for more spindle attachment sites and consequently for a "stronger" centromere [14], which provides preferential loss of maternal extra chromosome. At later postzygotic stages, while trisomy correction results in mosaicism for balanced translocation, preferable loss of maternal chromosome should not occur.

Sex-specific correction of transmitted translocation trisomy might explain either partly or entirely the phenomenon discussed since the 1960s, namely, transmission ratio distortion in offspring of female carriers of ROB [4–6]. Unfortunately, the precise mechanism of selective trisomy correction in female embryos is undefined.

3.3. Homologous Robertsonian translocations/isochromosomes: uneven involvement of acrocentric chromosomes, varying sex ratio, and no association with infertility

3.3.1. Rates and spectrum of HT in asymptomatic carriers

When groups of couples with reproductive disorders are compared (**Table 2**), tenfold difference is evident between them by both an incidence of HT carriers (0.03‰ in couples with infertility and 0.4% in couples with habitual abortion) and a proportion among all detected ROBs: 0.9% (1/111) with 95% CI of 0.2–4.9% vs. 10% (24/245) with CI of 7–14%, the difference is significant at $p < 0.0013$. And since the only carrier of HT in the group with infertility was a woman, one can assume that her "infertility" was due to early undiagnosed pregnancy losses.

In patients with male factor of infertility, it was originally intended to combine them with males from couples with infertility, especially since these groups did not statistically significantly differ either in the frequency of the detected ROB carriers (0.36 and 0.21‰, respectively) or in the spectrum of translocations. However, it was taken into account that in the surveyed couples, about half of males were partners of females with a female factor, and therefore their aggregation into one group is unnecessary. Nevertheless, despite the fact that in this group, the majority of the patients had a proven male infertility factor, proportion of HT carriers was only 3% (6/201 = 3.3 with 95% CI of 1.4–6.4%), which is not statistically different from that in the males from couples with infertility (0/91 = 0.0% with CI of 0.0–4%) at p = 0.18. Of note is that one of the six patients presented mosaicism for balanced/unbalanced HT [42].

Seventy-one single cases of HT carriers, including 48 females, were identified from the literature (Additional file S7). Almost all female carriers, except for two, were tested cytogenetically for multiple miscarriage and/or abnormal offspring. Of 23 male carriers, only 2 were tested for infertility, 1 of whom had mosaicism for an unbalanced rearrangement.

Table 5 presents the data collation from single reports, systematic surveys of couples with reproductive disorders, and also the publication of the authors who summarized the results of the diagnostic laboratory without detailing the indications for the testing. The most frequent were the HT of chromosome 13 and chromosome 22. A somewhat smaller number of asymptomatic carriers of HT of chromosomes 14 and 15 might be explained by the presence of imprinted genes on these chromosomes, a proportion of both HT14 and HT15 carriers have clinical manifestations depending on which of the parents the HT is inherited from (see Section 3.4).

The sex ratio in carriers of HT of chromosomes 13–15 and 21 is female biased, varying from 0.21 to 0.54, with the overall figure of 0.34 (22 M/64F) with 95% CI of 0.21–0.56. The predominance of female individuals among carriers of chromosome rearrangements of this type is explained by the sex-specific instability of pericentromeric regions [15, 69]. In contrast, sex

Translocations	Couple with reproductive disorders (Tables S5, S6)		Single cases tested for various reasons (Table S9)		Consecutive patients from a genetic unit [44]		Total		Sex ratio
	♂♂	♀♀	♂♂	♀♀	♂♂	♀♀	♂♂	♀♀	
13;13	2	6	1	15	2	3	5	24	0.21
14;14	1	4	5	6	1	3	7	13	0.54
15;15	2	1	3	9	0	2	5	12	0.42
21;21	1	1	4	8	0	6	5	15	0.33
22;22	3	4	10	8	2	1	15	13	1.15

Table 5. Spectrum of homologous translocations and sex ratio among carriers, updated from [43].

ratio among carriers of HT22 is not female biased (15 males/13 females, with 95% CI of 0.56–2.45), which might indicate some different "circumstances" of the formation of HT22 and the other acrocentric chromosomes. It is known that HT may have either a meiotic or mitotic origin and may be mono- or dicentric and biparental or uniparental [45]. All the information that the authors reported on the origin of HT is included in Additional file: Table S9. However, its scarcity does not allow drawing any conclusions as to the possible differences in the mechanisms of the formation of certain HT.

3.3.2. Problems of reproduction in carriers of HT

The data of the previous study suggested that homologous translocations do not contribute to a disturbance of spermatogenesis [8]. The present study showed that in patients with a male factor of infertility, the percentage of HT is 3% of the identified ROBs, in contrast to 10.5% in partners of women with miscarriage (although in the latter group about half of the individuals are partners of women with a female factor for infertility). It was noted that of the 22 male HT carriers (Additional file: Table S9), only 2 have been evaluated for infertility, 1 of them having a cell line with an unbalanced HT [3]. In the analysis of a testicular biopsy of another carrier, the authors found no reason to link the presence of HT with the impairment of his spermatogenesis [46].

Thus, in the overwhelming majority of cases, male HT carriers produce gametes capable of fertilization. The absence of spermatogenesis disorders, typical to nonhomologous ROB carriers, is most likely due to the ability of chromosome arms of HT to conjugate, as previously reported [47]. The authors, examining a man whose wives had habitual miscarriages, found completely normal spermogram parameters and testicular histology, wherein conjugation between the long arms of the isochromosome 14 took place in such a way that the chromosome did not differ from the usual bivalent. It is obvious that such a configuration is fraught with the possibility for formation of a ring chromosome. Indeed, in the offspring of two carriers of HT, there were children with ring chromosomes, most likely formed from parental HT [48, 49]. There are multiple reports in the literature on patients with ring chromosomes accompanying homologous translocations but of postzygotic origin [50–53]. Stetten et al. [53] suggested that the presence of HT is a necessary precursor to the formation of ring chromosomes.

Despite the fact that carriers of nonmosaic HT produce only abnormal gametes, there are cases of the birth of healthy children with the same rearrangement [54–59]. These rare cases can be the result of one of two mechanisms: the syngamy of a gamete carrying HT with a gamete nullisomic for the same chromosome or correction of a trisomic zygote by losing a free extra chromosome. It is curious that out of seven of these cases, in four of them, HT22 was transmitted. Studies of the inheritance events of balanced HTs provided initial evidence that chromosomes 13, 21, and 22 did not bear imprinted gene.

Several cases of the birth of healthy children with normal chromosomes to apparently nonmosaic HT carriers were reported [60–64]. The birth of chromosomally normal children indicates the presence of a normal line in the gonads of the parents with HT. In addition, one can assume a rare event—sporadic dissociation of centromere. This phenomenon was shown both for ROB [65, 66] and for nonacrocentric chromosomes [67, 68]. Another possibility was discussed as well, gonadal mosaicism in unbalanced HT (translocation trisomy), since gamete precursor cells with such a set of chromosomes are expected to produce 50% of daughter cells with normal karyotype [69].

It would seem that the feasibility of this possibility with respect to male patients is highly doubtful, since the presence of an additional chromosome induces spermatogenesis disorders. For example, it is well known that women with nonmosaic trisomy of chromosome 21 (Down's syndrome) are fertile, while men are mostly infertile, due to impaired spermatogenesis [70]. It is possible to assume that it is the presence of a cell line with unbalanced HT in the gonads as a result of incomplete correction of the original translocation trisomy that causes spermatogenesis disorders in carriers of apparently balanced HT.

Currently, infertility due to chromosomal abnormalities, with the corresponding pathologies of spermatogenesis, is overcome by reproductive technologies, and, paradoxically, it is possible that it is in male HT carriers with infertility that there is a chance to have a healthy offspring. For example, encouraging results were obtained using reproductive technologies for the production of healthy children from male carriers of trisomy 21 [71, 72].

In general, the reproductive prognosis for carriers of HT is pessimistic. But, given the nonzero chance of having gonadal mosaicism in them, we can recommend testing, the algorithm of which was published [69, 73]. In addition, another possibility of having a healthy child with the same rearrangement was discussed, that is, gamete donation from a carrier of the same balanced rearrangement, which does not carry imprinted genes [73].

3.4. Sex ratio in ill-defined carriers of homologous translocations/isochromosomes

A scrupulous search in available literature yielded 10 ill-defined carriers of HT14 and 28 carriers of HT15 (Additional file: S10). Although the number of published cases of HT with clinical manifestation of uniparental disomy is small, there are some observations of interest.

3.4.1. Sex ratio in patients with UPD(HT14)

Unlike asymptomatic individuals with biparental HT14, patients with UPD(HT14) demonstrate some male predominance (6 M/2F), while the majority of them (eight of ten) had maternally derived rearrangement. More cases are needed for solid conclusion on the SR in this group.

3.4.2. Sex ratio in patients with maternal UPD(HT15), Prader-Willi syndrome

Strong female predominance among patients with maternal UPD(HT15) was first reported in the discussion of the concept of trisomy correction due to parent-sex-specific loss [15]. In previous studies, a male predominance among patients with maternal non-ROB UPD (15) was suggested to be the result of either a bias of ascertainment due to milder phenotype in female UPD patients or difference in survival of early trisomy 15 conceptuses [74]. However, in contrast, Kovaleva noted that among patients with UPD(HT15), there was no male predominance, with five male and ten female carriers [15]. Mitchel et al. also suggested a possible difference in the probability of trisomic zygote rescue depending on the sex [74]. However, the predominant rescue of trisomic male zygotes would result in a male predominance in mosaic cases, while no male predominance was reported in a collective sample of 50 fetuses with T15 mosaicism (SR = 0.67) [15]. Kovaleva suggested that the male prevalence among patients with non-ROB UPD(15) can be explained by female-specific loss of a maternal chromosome, causing biparental inheritance and therefore complete correction of trisomy in females (without UPD) [15]. For an explanation of the female predominance among carriers of UPD(HT15), parent-sex-specific loss should be considered, but in this case, a preferential loss of paternal extra chromosome from female trisomic zygotes with unbalanced HT is suggested.

3.4.3. *Sex ratio in patients with paternal UPD(HT15), Angelman syndrome*

Nine reported HT15 carriers with Angelman syndrome were males. All of eight tested for UPD patients had paternal isodisomy. Among homologous HT, the majority of them were established to be isochromosomes. Several mechanisms of isochromosomes formation were discussed, including gametic complementation, trisomy rescue, and monosomy rescue. It was suggested that they mainly should be formed postzygotically (see for review [73]). However, postzygotic formation of pericentromeric rearrangements is essentially female-specific [15, 69].

A strong male prevalence among patients with UPD(HT15) can be explained by meiotic event, nonhomologous co-orientation of the isochromosome with X chromosome during the first meiotic division in the spermatocyte. In such a case, X chromosome and isochromosome travel to the opposite poles, providing preferential segregation of isochromosome with Y chromosome. This mechanism, proven for *Drosophila* [75, 76], was proposed to explain male excess among carriers of paternally derived regular trisomy 21 [77], as well as male-biased SR in trisomic offspring fathered by carriers of dup(21) [78], and in trisomy 21 offspring inherited paternal noncontributing rearrangement [79].

4. Conclusion

It is interesting that very recently the epidemiology of Robertson translocations was suggested to this author as not worthy of any attention. Currently, in this field there are multiple unanswered questions. Further studies are required to elucidate the nature of female preponderance among carriers of Robertsonian translocation in newborns, as well as of other intriguing phenomena uncovered in this paper, such as a nonuniformity in the HT spectrum and difference in sex ratio between the carriers of the HT22 and the carriers of HT of the other acrocentric chromosomes. Moreover, chromosome 22 is rather mysterious in the context of the differences in the spectrum of nonhomologous translocations between groups of patients with reproductive disorders. There is no clear understanding of the role of HT in the etiology of male infertility and what factors determine the association of part of HT with impaired spermatogenesis. In addition, there are some aspects of ROB epidemiology not considered in this chapter, including interchromosomal effect and mosaicism.

Acknowledgements

The author is greatly indebted to Prof. Philip D. Cotter (University of California, San Francisco, USA) for the helpful comments and amending English in this paper and to Dr. Nikita N. Khromov-Borisov (Almazov National Medical Research Centre, St. Petersburg, Russia) for statistical analysis of the data.

Author details

Natalia V. Kovaleva

Address all correspondence to: kovalevanv2007@yandex.ru

Academy of Molecular Medicine, St. Petersburg, Russian Federation

References

[1] Gada Saxena S, Desai K, Shawale L, et al. Chromosomal aberrations in 2000 couples of Indian ethnicity with reproductive failure. Reproductive Biomedicine Online. 2012;**25**(2): 209-218. DOI: 10.1016/j.rbmo.2012.04.004

[2] Mau UA, Bäckert IT, Kaiser P, Kiesel L. Chromosomal findings in 150 couples referred for genetic counselling prior to intracytoplasmic sperm injection. Human Reproduction. 1997; **12**(5):930-937

[3] Veld PA, Weber RF, Los FJ, et al. Two cases of Robertsonian translocations in oligozoospermic males and their consequences for pregnancies induced by intracytoplasmic sperm injection. Human Reproduction. 1997;**12**(8):1642-1644

[4] Hamerton JL. Robertsonian Translocations in Man: Evidence on Segregation from Family Studies. Pfizer Medical Monographs no. 5. Edinburgh: University of Edinburgh Press; 1970

[5] Boue A, Gallano P. A collaborative study of the segregation of inherited chromosome structural rearrangements in 1356 prenatal diagnoses. Prenatal Diagnosis. 1984;**4**(Specno):45-67

[6] De Villena FP-M, Sapienza C. Transmission ratio distortion in offspring of heterozygous female carriers of Robertsonian translocations. Human Genetics. 2001, n.d.;**108**(1):31-36

[7] Shaffer LG. Risk estimates for uniparental disomy following prenatal detection of a nonhomologous Robertsonian translocation. Prenatal Diagnosis. 2006;**26**(4):303-307. DOI: 10.1002/pd.1384

[8] Kovaleva NV. Examination of rates and spectrums of Robertsonian translocations in the general population and in patients with reproductive disorders. Russian Journal of Genetics. 2018;**54**(4):489-493. DOI: 10.1134/S1022795418040099

[9] Harrison JC, Schwab C. Constitutional abnormalities of chromosome 21 predispose to IAMP21-acute lymphoblastic leukaemia. European Journal of Human Genetics. 2016; **59**(3):162-165. DOI: 10.1016/j.ejmg.2016.01.006

[10] De Gregori M, Ciccone R, Magini P, et al. Cryptic deletions are a common finding in "balanced" reciprocal and complex rearrangements: A study of 59 patients. Journal of Medical Genetics. 2007;**44**(12):750-762. DOI: 10.1136/jmg.2007.052787

[11] Höckner M, Spreiz A, Frühmesser A, et al. Parental origin of de novo cytogenetically balanced reciprocal non-Robertsonian translocations. Cytogenetic and Genome Research. 2012;**136**(4):242-245. DOI: 10.1159/000337923

[12] Page SL, Shaffer LG. Chromosome stability is maintained by short intercentromeric distance in functionally dicentric human Robertsonian translocations. Chromosome Research. 1998;**6**(2):115-122

[13] Bandiopadhyay R, Heller A, Knox-DuBois C. Parental origin and timing of de novo Robertsonian translocation formation. American Journal of Human Genetics. 2002;**71**(6): 1456-1462. DOI: 10.1086/344662

[14] Daniel A. Distortion of female meiotic segregation and reduced male fertility in human Robertsonian translocations: Consistent with the centromere model of co-evolving centromere DNA/centromeric histone (CENP-A). American Journal of Medical Genetics. 2002; **112**(4):450-452. DOI: 10.1002/ajmg.10618

[15] Kovaleva NV. Sex-specific chromosome instability in early human development. American Journal of Medical Genetics. 2005;**136A**(1):401-413. DOI: 10.1002/ajmg.a.30815

[16] Kovaleva NV. Germ-line transmission of trisomy 21: Data from 80 families suggest an implication of grandmaternal age and a high frequency of female-specific trisomy rescue. Molecular Cytogenetics. 2010;**3**:7. DOI: 10.1186/1755-8166-3-7

[17] Ferguson-Smith MA, Yates JRW. Maternal age-specific rates for chromosome aberrations and factors influencing them: Report of a collaborative European study on 52 965 amniocenteses. Prenatal Diagnosis. 1984;**4**(4, Special issue):5-45

[18] Chen C-P, Chern S-R, Wu P-C, et al. Unbalanced and balanced acrocentric rearrangements involving chromosomes other than chromosome 21 at amniocentesis. Taiwanese Journal of Obstetrics & Gynecology. 2009;**48**(4):389-399. DOI: 10.1016/S10284559(09)60329-6

[19] Kovaleva NV. An overlooked phenomenon: Female-biased sex ratio among carriers of Robertsonian translocations detected in consecutive newborn studies. Russian Journal of Genetics. 2017;**53**(12):1366-1373. DOI: 10.1134/S1022795417120067

[20] Antonarakis SE, Blouin JL, Maher J, et al. Maternal uniparental disomy for human chromosome 14 due to loss of a chromosome 14 from somatic cells with t(13;14) trisomy. American Journal of Medical Genetics. 1993;**52**(6):1145-1115

[21] Barton DE, McQuaid S, Stallings R. Further evidence for an emerging maternal uniparental disomy chromosome 14 syndrome: Analysis of a phenotypically abnormal de novo

obertsonian translocation t(13;14) carrier. American Journal of Human Genetics. 1996;**59**(Suppl):687

[22] Coviello DA, Panucci E, Manttero MM, et al. Maternal uniparental disomy for chromosome 14. Acta Geneticae Medicae et Gemellologiae. 1996;**45**(1-2):169-172

[23] Desilets VA, Yong SI, Kalousek DK, et al. Maternal uniparental disomy for chromosome 14. American Journal of Human Genetics. 1997;**61**(Suppl):691

[24] Giunti L, Lapi E, Guarducci S, et al. Maternal heterosomy for chromosome 14 and 13/14 Robertsonian translocation in a female with normal development, short stature, and dysmorphic features. European Journal of Human Genetics. 2002;**10**(Suppl 1):120

[25] Healey S, Powell F, Battersby M, et al. Distinct phenotype in maternal uniparental disomy of chromosome 14. American Journal of Medical Genetics. 1994;**51**(2):147-149. DOI: 10.1002/ajmg.1320510213

[26] Mitter D, Buiting K, von Eggeling F, et al. Is there a higher incidence of maternal uniparental disomy 14 [upd(14) mat]? Detection of 10 new patients by methylation-specific PCR. American Journal of Medical Genetics. 2006;**140A**(19):2039-2049. DOI: 10.1002/ajmg.a.31414

[27] Takanashi I, Takanashi T, Utsanomiya M, et al. Long-acting gonadotropin-releasing hormone analogue treatment for central precocious puberty in maternal disomy chromosome 14. The Tohoku Journal of Experimental Medicine. 2005;**207**(4):333-338. DOI: 10.1620/tjem.207.333

[28] Harrison KJ, Allingham-Hawkins DJ, Hummel J, et al. Risk of uniparental disomy in Robersonian translocation carriers: Identification of upd(14) in a small cohort. American Journal of Human Genetics. 1998;**63**(Suppl 4):51

[29] Link L, McMillin K, Popovich B, Magenis RE. Maternal uniparental disomy for chromosome 14. American Journal of Human Genetics. 1996;**59**(Suppl):687

[30] Temple IK, Cockwell A, Hassold T, et al. Maternal uniparental disomy for chromosome 14. Journal of Medical Genetics. 1991;**28**(8):511-514

[31] Worley KA, Rundus VR, Lee EB, et al. Maternal uniparental disomy 14 presenting as language delay. American Journal of Human Genetics. 2001;**69**(Suppl 4):738

[32] Cotter PD, Kaffe S, McCurdy LD, et al. Paternal uniparental disomy for chromosome 4: A case report and review. American Journal of Medical Genetics. 1997;**70**(1):74-79

[33] Kurosawa K, Sasaki H, Sato Y, et al. Paternal UPD14 is responsible for a distinctive malformation complex. American Journal of Medical Genetics. 2002;**110**(3):268-272. DOI: 10.1002/ajmg.10404

[34] Wang J-CC, Passage MB, Yeh PH, et al. Uniparental heterosomy for chromosome 14 in a phenotypically abnormal familial balanced 13/14 Robertsonian translocation carrier. American Journal of Human Genetics. 1991;**48**(6):1069-1074

[35] Yano S, Li L, Owen S, et al. Further delineation of the paternal uniparental disomy (UPD) 14: The fifth reported liveborn case. American Journal of Human Genetics. 2001;**69**(Suppl 4):739

[36] Senci A, Cavani S, Villa N, et al. Nonhomologous Robertsonian translocations and uniparental disomy risk: An Italian multicentric prenatal survey. Prenatal Diagnosis. 2004;**24**(8): 647-652. DOI: 10.1002/pd.962

[37] Smith KK, Boyle TA, Morgan DL, Parkin CA. Uniparental disomy: UK Collaborative Study. American Journal of Human Genetics. 2001;**69**(Suppl):911

[38] Berends JW, Hordijk R, Oosterwijk JC, et al. Two cases of maternal uniparental disomy 14 with a phenotype overlapping with the Prader-Willi phenotype. American Journal of Medical Genetics. 1999;**84**(1):76-79

[39] Ruggeri A, Dulcetti F, Miozzo M, et al. Prenatal search for UPD 14 and UPD15 in 83 cases of familial and de novo heterologous Robertsonian translocations. Prenatal Diagnosis. 2004;**24**(12):997-1000. DOI: 10.1002/pd.961

[40] Barbash-Hazan S, Frumkin T, Malcov M, et al. Preimplantation aneuploid embryos undergo self-correction in correlation with their developmental potential. Fertility and Sterility. 2009;**92**(3):890-895. DOI: 10.1016/j.fertnstert.2008.07.1761

[41] Bazrgara M, Gourabia H, Valojerdic MR, et al. Self-correction of chromosomal abnormalities in human preimplantation embryos and embryonic stem cells. Stem Cells and Development. 2013;**22**(17):2449-2456. DOI: 10.1089/scd.2013.0053

[42] Tuerlings JHAM, de France HF, Hamers A, et al. Chromosome studies in 1792 males prior to intra—Cytoplasmic sperm injection: The Dutch experience. European Journal of Human Genetics. 1998;**6**(3):194-200. DOI: 10.1038/sj.ejhg.5200193

[43] Zhao W-W, Wu M, Chen F, et al. Robertsonian translocations: An overview of 872 Robertsonian translocations identified in a diagnostic laboratory in China. PLoS One. 2015;**10**(5):e0122647. DOI: 10.1371/journal.pone.0122647

[44] Kovaleva NV. Homologous Robertsonian translocations: Spectrum, sex ratios, reproductive risks. Russian Journal of Genetics. 2018;**54**. in press

[45] Robinson WP, Bernasconi F, Basaran S, et al. A somatic origin of homologous Robertsonian translocations and isochromosomes. American Journal of Human Genetics. 1994;**54**(2): 290-302

[46] Laurent C, Papathanassiou Z, Haour P, Cognat M. Mitotic and meiotic studies on 70 cases of male sterility. Andrologie. 1973;**5**(3):193-200

[47] Hulten M, Lindsten J. The behavior of structural aberrations at male meiosis. In: Jacobs PA, Price WH, Law P, editors. Human Population Cytogenetics. Edinburgh, United Kingdom: Edinburgh University Press. pp. 23-61

[48] de Almeida JCC, Llerena JC Jr, Gomes DM. Ring 13 in an adult male with a 13;13 translocation mother. Annales de Génétique. 1983;**26**(2):112-115

[49] Neri G, Ricchi R, Pelino A, et al. A boy with ring chromosome 15 derived from a t(15q;15q) Robertsonian translocation in the mother: Cytogenetic and biochemical findings. American Journal of Medical Genetics. 1983;**14**(2):307-314. DOI: 10.1002/ajmg.1320140211

[50] Adam LR, Kashork CD, Van den Veyver IB, et al. Ring chromosome 15: Discordant karyotypes in amniotic fluid, placenta and cord. American Journal of Human Genetics. 1998;**63**(Suppl):A126

[51] Dallapiccola B, Bianco I, Brinchi V, et al. t(21q;21q)/r(t(21q;21q)) mosaic in two unrelated patients with mild stigmata of Down syndrome. Annales de Génétique. 1982;**25**(1):56-58

[52] Pangalos C, Vellisariou V, Ghica M, Liacacos D. Ring-14 and trisomy 14q in the same child. Annales de Génétique. 1984:**27**(1)

[53] Stetten G, Tuck-Miller C, Blakemore KJ, et al. Evidence for involvement of a Robertsonian translocation 13 chromosome in formation of a ring chromosome 13. Molecular Biology & Medicine. 1990;**7**(6):479-484

[54] Borgaonkar DS. Repository of human chromosomal variants and anomalies. 13th ed. Newark, Delaware: Medical Center of Delaware; 1999. p. 352

[55] Chopade DK, Harde H, Ugale P, Chopade S. Unexpected inheritance of a balanced homologous translocation t(22q;22q) from father to a phenotypically normal daughter. Indian Journal of Human Genetics. 2014;**20**(1):85-89. DOI: 10.4103/0971-6866.132765

[56] Kirkelis VGHJ, Hustinx TWJ, Scheres JMJC. Habitual abortion and translocation (22q; 22q): Unexpected transmission from a mother to her phenotypically normal daughter. Clinical Genetics. 1980;**18**(6):456-461. DOI: 10.1111/j.1399-0004.1980.tb01794.x

[57] Miny P, Koppers B, Bogdanova N, et al. Paternal uniparental disomy 22. American Journal of Medical Genetics. 1995;**7**(Suppl):216

[58] Palmer CG, Schwartz S, Hodes ME. Transmission of a balanced homologous t(22q;22q) translocation from mother to normal daughter. Clinical Genetics. 1980;**17**(6):418-422

[59] Slater H, Shaw JH, Dawson G, et al. Maternal uniparental disomy of chromosome 13 in a phenotypically normal child. Journal of Medical Genetics. 1994;**31**(8):644-646

[60] Cinar C, Beyazyurek C, Ekmekci CG, et al. Sperm fluorescence in situ hybridization analysis revels normal sperm cells for 14;14 homologous male Robertsonian translocation carrier. Fertility and Sterility. 2011;**95**(1):e285-e289. DOI: 10.1016/j.fertnstert.2010.05.033

[61] Daniel A, Hook EB, Wulf G. Risks of unbalanced progeny at amniocentesis to carriers of chromosome rearrangements: Data from United States and Canadian laboratories. American Journal of Medical Genetics. 1989;**33**(1):14-53. DOI: 10.1002/ajmg.1320330105

[62] Lipson MH, Breg WR. Non-karyotyping evidence for mosaicism in 15;15 translocation: Implications for genetic counseling and patient management. American Journal of Human Genetics. 1978;**30**(Suppl 6):58A

[63] Lucas M. Translocation between both members of chromosome pair number 15 causing recurrent abortions. Annals of Human Genetics. 1969;**32**(4):347-352

[64] Van Erp F. Offspring of a male 45,XY,der(22;22)(q10;q10) carrier. European Journal of Human Genetics. 2016;**24**(Suppl 1):58

[65] Fujimoto A, Lin MS, Korula SR, Wilson MG. Trisomy 14 mosaicism with t(14;15)(q11;p11) in offspring of a balanced translocation carrier mother. American Journal of Medical Genetics. 1985;**22**(2):333-342. DOI: 10.1002/ajmg.1320220217

[66] McFadden DE, Dill F, Kalousek DK. Fission in 1q isochromosome. American Journal of Human Genetics. 1986;**39**(Suppl 3):A133

[67] Fryns JP, Kleczkowska A, Limbos C, et al. Centric fission of chromosome 7 with 47,XX,del (7)(pter->cen::q21->qter)+cen fr karyotype in a mother and proximal 7q deletion in two malformed newborns. Annales de Génétique. 1985;**28**(4):248-250

[68] Del Porto G, Di Fusco C, Baldi M, et al. Familial centric fission of chromosome 4. Journal of Medical Genetics. 1984;**21**(5):388-391

[69] Kovaleva NV. Nonmosaic balanced homologous translocations of major clinical significance: Some may be mosaic. American Journal of Medical Genetics. 2007;**143A**(23):2843-2850. DOI: 10.1002/ajmg.a.31745

[70] Hsiang YH, Berkovitz GD, Bland GL, et al. Gonadal function in patients with Down syndrome. American Journal of Medical Genetics. 1987;**27**(2):449-458. DOI: 10.1002/ajmg.1320270223

[71] Kim ST, Cha YB, Park JM, Gye MC. Successful pregnancy and delivery from frozen thawed embryos after cytoplasmic sperm injection using round-headed spermatozoa and assisted oocyte activation in a globozoospermic patient in mosaic Down syndrome. Fertility and Sterility. 2001;**75**(2):445-447. DOI: 10.1016/S0015-0282(00)01698-8

[72] Aghajanova L, Popwell JM, Chetkowski RJ, Herndon CN. Birth of a healthy child after preimplantation genetic screening of embryos from sperm of a man with non-mosaic Down syndrome. Journal of Assisted Reproduction and Genetics. 2015;**32**(9):1409-1413. DOI: 10.1007/s10815-015-0525-z

[73] Kovaleva NV, Shaffer LG. Under-ascertainment of mosaic carriers of balanced homologous acrocentric translocations and isochromosomes. American Journal of Medical Genetics. 2003;**121A**(2):180-187. DOI: 10.1002/ajmg.a.20156

[74] Mitchell J, Schinzel A, Langlois S, et al. Comparison of phenotype in uniparental disomy and deletion Prader-Willi syndrome: Sex specific differences. American Journal of Medical Genetics. 1996;**65**(2):133-136. DOI: 10.1002/(SICI)1096-8628(19961016)65:2<133::AIDAJMG10>3.0.CO;2-R

[75] Grell RF. Distributive pairing in man? Annales de Génétique. 1971;**14**(3):165-171

[76] Chadov BF. From the phenomenon of nondisjunction to the problem of chromosome co orientation (75th anniversary of Bridges' article). Genetika. 1991;**27**(11):1877-1903

[77] Kovaleva NV. Distributive pairing of chromosomes and aneuploidy in man. Genetika. 1992;**28**(10):5-15

[78] Kovaleva NV. Additional evidence for nonhomologous meiotic co-orientation (NMC) in man. Chromosome Research. 2005;**13**(Suppl 1):41

[79] Kovaleva NV. Trisomy 21 in offspring of carriers of balanced non-contributing autosomal rearrangement. Examination of interchromosomal effect and nonhomologous meiotic co-orientation. In: van den Bosch A, Dubois E, editors. New Developments in Down Syndrome Research. NY: Nova Science Publishers, Inc.; 2012. pp. 149-176. Available from: https://www.novapublishers.com/catalog/product_info.php?products_id=376

Chapter 7

Comparative Cytogenetics Allows the Reconstruction of Human Chromosome History: The Case of Human Chromosome 13

Rita Scardino, Vanessa Milioto and
Francesca Dumas

Additional information is available at the end of the chapter

http://dx.doi.org/10.5772/intechopen.79380

Abstract

Comparative cytogenetics permits the identification of human chromosomal homologies and rearrangements between species, allowing the reconstruction of the history of each human chromosome. The aim of this work is to review evolutionary aspects regarding human chromosome 13. Classic and molecular cytogenetics using comparative banding, chromosome painting, and bacterial artificial chromosome (BAC) mapping can help us formulate hypotheses about chromosome ancestral forms; more recently, sequence data have been integrated as well. Although it has been previously shown to be conserved when compared to the ancestral primate chromosome, it shows a degree of rearrangements in some primate taxa; furthermore, it has been hypothesised to have a complex origin in eutherian mammals which has still not been completely clarified.

Keywords: FISH, evolution, mammals, human synteny

1. Introduction

Comparative cytogenetics has been widely applied to many mammalian species [1–3] through banding methods and, later, with fluorescence *in situ* hybridization (FISH) of whole chromosomes and bacterial artificial chromosome (BAC) probes; these approaches permit the definition of regions of chromosomal homology, rearrangements, and breakpoints, as well as elucidate phylogenetic relationships between taxa [4]. In addition, the comparative cytogenetic approach is particularly useful in the reconstruction of human chromosome (HSA) history.

Indeed, parsimony analysis of homologies and rearrangements permits us to define ancestral chromosomal syntenies (synteny is the colocalization of two or more genetic loci) and derived ones [2]. Banding allows us to first evaluate rearrangements between species; the mapping of whole chromosomes through the chromosomal painting approach allows researchers to better define rearrangements at the molecular level, such as Robertsonian ones and breakpoints. At a finer level, the use of DNA cloned inside vectors such as yeast artificial chromosomes (YACs) or BAC, used as mapping probes, permits the evaluation of chromosomal dynamics [5, 6], defining marker orders and intrachromosome rearrangements. Moreover, the use of specific loci or repetitive probes permits the localization of specific sequences, such as repetitive ones, which are often supposed to be responsible for the plasticity of chromosomes [7–10] and human genes involved in cancers [11].

More recently, the integration of cytogenetic data with sequence data has been proposed [12–16]. These kinds of data are available from genomic browsers and are helpful for testing previously proposed phylogenomic hypotheses and chromosomal organisation reconstructions.

In this review, we report the principal approach which has proven useful for studying human chromosome history by analysing previous cytogenetic and sequence data regarding human chromosome 13.

2. The reconstruction of human chromosome history

At least three or four principal approaches can be used to reconstruct human synteny history. In a comparative perspective, the analysis of banding data permits the identification of chromosomal homologies. In particular, the analysis of the banding patterns obtained by the enzymatic digestion of chromosomes in metaphases using proteolysis and Giemsa solution staining permits the identification of chromosomal homologies and principal rearrangements occurring between species. Consequently, by focusing attention on a single chromosome, it is possible to track the principal evolutionary steps involving each individual human chromosome [1].

Another approach is the analysis of comparative painting data; the painting approach consists of a whole chromosome undergoing FISH on cytogenetic preparations, allowing the identification of molecular level homologies, interchromosomal rearrangements and genomic breakpoints. First, human chromosome probes are mapped onto metaphases of target species (chromosome painting [CP]) [17], then, for a better comparison, animal chromosomal probes are mapped onto human metaphases in a reciprocal hybridization (RP) [18]. Subsequently, whole animal chromosomes are mapped onto other animal metaphases in an approach known as ZOO-FISH, Z-F [19]. The analysis of these data regarding a single chromosome, consequentially, permits the tracking of each change involving the human chromosome under study.

In addition, human chromosome evolution can be studied using another kind of probe, the BAC probe, containing an insert of 50–300 Kb of the human genome. It can be mapped by FISH onto the metaphases of many species. BACs are available for each human chromosome and can be purchased from the BAC/PAC Resource Center (Chori), and some of them are commercially available for medical diagnosis. These probes are very useful in detecting small interchromosomal rearrangements which are not detectable by painting and in defining marker order along chromosomes, thus revealing inversions, new centromere evolutions (new centromeres arise without the occurrence of inversions, maintaining the marker order), and duplications [3].

Comparative cytogenetics has been applied to reconstructing most human chromosome history; these published works have mainly been done by reviewing previous painting data or by mapping BAC probes on primates (see review in [2]) and other eutherian mammals; some works have analysed only specific chromosome regions (see **Table 1** for representative works).

Furthermore, alignments of sequences (SA) of many mammal species, obtainable from the NCBI, UCSC, and Ensemble genome browsers, can be integrated with molecular cytogenetic information in order to shed light on the history and peculiar features characterising each human chromosome.

2.1. The evolutionary history of HSA 13

Human chromosome 13 has been sequenced, and it has been shown to be the largest acrocentric chromosome in the human karyotype. Currently, the NCBI reports 1381 total genes, 41 novel genes, and 477 pseudogenes for a size of 114.36 MB [47]. It is among the human chromosomes with the lowest percentage of duplicated sequences [48].

The analyses of classical and molecular cytogenetics, using comparative banding and chromosome painting, have allowed researchers to formulate hypotheses about its ancestral forms. In this report, we delineate the principal steps regarding the history of human chromosome 13, tracked through the analysis of previous cytogenetics literature and sequence data. We have reported a list of species analysed by painting or sequence information, chromosome homologues to human chromosome 13, human associations with HSA 13, chromosome type if available, references and methods from which we obtained the data, such as CP, RP, Z-F, and SA (see **Table 2**). The principal steps in the evolution of human chromosome 13 are illustrated in a graphical reconstruction of the mammal phylogenetic tree, **Figure 1**; the mammal phylogenetic tree has been drawn in agreement with previous ones [16, 49], with some modifications, and was created using Mesquite v.2.75 [50]. Among mammals, three major groups are distinguishable: monotremes (Prototheria, platypus), marsupials (Metatheria, opossum), and placental mammals (eutherian), with these last two known as Theria; among placental mammals, Afrotheria, Xenarthra, and Boroautherian are recognized, with the latter comprising Laurasiatheria and Euarchontoglires (or Supraprimates) [49]. In the mammalian phylogenetic tree are shown the orthologue blocks that correspond to human chromosome 13—in yellow—in representative eutherian species for which reciprocal chromosome painting is available; for some of them also DNA sequence alignments have been previously showed, see **Table 2** for reference. For each species are reported chromosome ideograms on which human synteny 13 is found, and on the left of the ideograms are reported the species' chromosome number and on the right HSA syntenies; the black circle is the centromere. Syntenies homologues of human chromosome 13 in platypus (Monotremata) are on chromosomes 2, 10, and 20, in opossum (Metatheria) are on chromosomes 4 and 7, and in chicken (Aves) are on chromosome 1. These chromosomes are reported in box because they are representative eutherian mammal

HSA chr.	Methods	References
1	Region study by BAC mapping	[20]
	History by multidisciplinary approach	[21]
	History by BAC mapping	[2]
2	Region study by BAC mapping	[22]
	History by BAC mapping	[2]
3	Region study by BAC mapping	[23–25]
	Review	[26]
4	Region study by BAC mapping	[27]
	History by BAC mapping	[2]
	Region study by BAC mapping	[28, 29]
5	Region study by BAC mapping	[30, 31]
	review	[2]
6	History by BAC mapping	[32, 33]
7	Painting	[34]
	Review	[35]
	Region study by BAC mapping	[36]
8	Brief history by BAC mapping	[2]
9	Region study by BAC mapping	[37]
10	History by BAC mapping	[37, 38]
11	History by BAC mapping	[39]
12	Brief history by BAC mapping	[2]
13	History by BAC mapping	[40]
14	Region study by BAC mapping	[41]
15	Region study by BAC mapping	[41, 42]
16	History by BAC mapping,	[43]
	Painting	[34]
17	History by BAC mapping	[2]
18	Region study by BAC mapping,	[44]
	History by BAC mapping	[2]
19	Painting,	[34]
	Brief history by BAC mapping	[2]
20	History by BAC mapping	[45]
21	Region study by BAC mapping	[23]
	Brief history by BAC mapping	[2]
22	Brief history by BAC mapping	[2]
X	Brief history by BAC mapping	[2]
y	Region study by BAC mapping	[46]

Table 1. List of representative works, (references and methods) analyzing each human chromosome evolution and/or marker order in particular chromosomal region.

	Chromosome type	Chr.	Human association	References	Methods
Dermoptera					
Galeopterus variegatus	Acrocentric	13		[58]	RP
Proboscidea					
Loxodonta africana	Acrocentric	16,26	13, 6/13/3	[59]	CP
	Submetacentric			[12, 16]	SA
Elephas maximus	Acrocentric	16, 26	13, 6/13/3	[59]	CP
	Submetacentric				
Tubulidentata					
Orycteropus afer	Submetacentric	1	19/16/13/2/8/4	[59, 60]	CP
					SA
Afrosoricida					
Chrysochloris asiatica	Metacentric	8	13/18	[61, 60]	RP
					SA
Macroscelidea					
Elephantulus rupestris	Submetacentric	2	13/3/21/5	[61]	CP
Elephantulus edwardii				[60]	SA
Macroscidelis proboscidens	Submetacentric	2	13/3/21/5	[53]	CP
Sirenia					
Trichechus manatus	Metacentric	19	13/3	[62]	CP
Eulipotyphla					
Sorex araneus	Metacentric	bc	9/5/2/13/8/7	[16, 63]	CP, SA
Blarinella griselda	Submetacentric	3	13/10/13/4/5	[63]	CP
Neotetracus sinensis	Submetacentric	3,10	13/4/20/10,	[63]	CP
	Acrocentric		1/13/10/12/22		
Hemiechinus auritus				[64]	CP
Talpa europaea	Metacentric	6	2/13	[65]	CP
Cingulata					
Dasypus novemcinctus	Submetacentric	19		[66]	CP
Pilosa					
Choloepus didactylus	Acrocentric	17		[64]	CP
Coniochaeta hoffmannii	Acrocentric	12		[66]	CP
Tamandua tetradactyla	Metacentric	4, (2*)	13/1	[64, *66]	CP
Bradypus torquatus	Acrocentric	12		[67]	CP
Bradypus variegatus	Acrocentric	17		[67]	CP
Carnivora					
Mustela putorius				[68]	CP

	Chromosome type	Chr.	Human association	References	Methods
Vulpes vulpes	Submetacentrics	6,9	13/14, 2/8/13/3/19	[69]	RP
Canis lupus familiaris	Acrocentrics	(25*) 22, 28		[*70]	RP
				[69]	CP
				[71, 72]	Z-F
				[16]	SA
Felis silvestris catus	Acrocentric	A1		[69]	CP
				[12, 13, 51]	SA
Mephitis mephitis	Submetacentric	19		[73]	CP
Procyon lotor	Metacentric	3	13/2	[73]	CP
Perissodactyla					
Equus caballus	Acrocentric	17		[74]	RP
				[13, 16]	SA
				[19]	Z-F
				[40]	BAC
Equus asinus		11		[19]	Z-F
Equus burchelli	Submetacentric	6q	13/9	[19]	RP
					Z-F
Equus grevyi		6q	13/9	[19]	Z-F
Equus zebra hartmannae		15		[19]	Z-F
Equus hemionus onager		5q	12/13/22	[19]	Z-F
Equus przewalskii		16		[19]	Z-F
Diceros bicornis	Acrocentric	10		[19]	Z-F
Ceratotherium simum		10		[19]	Z-F
Tapirus bairdii		1		[19]	Z-F
Tapirus indicus	Acrocentric	18		[19]	Z-F
Tapirus pinchaque		13		[19]	Z-F
Tapirus terrestris		8		[19]	Z-F
Hemiechinus auritus	Submetacentrics	5q,6	5/13, 2/22/12/13/12	[64]	CP
Pholidota					
Manis javanica	Submetacentric	1,9q	13/5/2p, 18/13	[64]	CP
	Metacentric			[75]	CP
Manis pentadactyla	Submetacentric	1q, 17	13/5/2, 13	[75]	CP
	Acrocentric				
Cetartiodactyla					
Bos taurus	Acrocentric	12		[12, 16]	SA
				[76]	RP

	Chromosome type	Chr.	Human association	References	Methods
Sus scrofa	Metacentric	11		[12, 16]	SA
				[76]	RP
Camelus dromedarius	Metacentric	14		[76]	RP
Globicephala melas	Metacentric	15		[77]	Z-F
Hippopotamus amphibious	Metacentric	15		[77]	Z-F
Giraffa camelopardalis	Metacentric	12	14/15/13	[77]	Z-F
Okapia johnstoni	Acrocentric	11		[77]	Z-F
Moschus moschiferus	Acrocentric	17		[77]	Z-F
Lagomorpha					
Oryctolagus cuniculus	Submetacentric	8	13/12	[78]	RP
				[51]	SA
Rodentia					
Mus musculus		3,5,8,14,14		[13]	SA
				[79]	SA
				[16]	SA
Rattus norvegicus		2,12,15,15,16		[13]	SA
				[16]	SA
Sciurus carolinensis	Submetacentric	6	10/13	[80]	RP
				[81]	RP
Petaurista albiventer	Metacentric	11	10/13	[81]	CP
Tamias sibiricus	Metacentric	10	10/13	[81]	CP
Castor fibre	Submetacentric	4	8/13	[79]	CP
Pedetes capensis	Submetacentric	6	13/12/22	[79]	CP
Sicista betulina	Metacentric, Submetacentric	1,9	13/4/10/11/9/10, 3/6/313/19	[79]	CP
Chiroptera					
Eonycteris spelaea	Submetacentric	E11	13/4/8/13	[82]	CP
Rhinolophus mehelyi	Acrocentric	R6	13/4/8/13	[82]	CP
Hipposideros larvatus	Metacentric	H1	13/3/21	[82, 83]	CP
Mormopterus planiceps	Metacentric	M7	13/18	[82]	CP
Myotis myotis	Metacentric	V5/6	4/8/13/12/22	[82]	CP
Aselliscus stoliczkanus	Metacentric	1	22/12/13/4/8/13	[83]	CP
Megaderma spasma	Metacentric	12	20/13/8b/4c	[84]	CP
Taphozous melanopogon	Submetacentric	1	4c/8b/13/16b/7c/5a	[84]	CP

	Chromosome type	Chr.	Human association	References	Methods
Primates					
Strepsirrhini					
Avahi laniger		12		[85]	CP
Daubentonia madagascariensis		8p	10/13	[85]	CP
Eulemur fulvus		12		[85]	CP
Hapalemur griseus griseus		15		[85]	CP
Indri indri	Submetacentric	3p	13/17	[85]	CP
Lemur catta	Acrocentric	13		[85, 86]	BAC
					CP
Lepilemur ankaranensis		14		[87]	CP
Lepilemur dorsalis		6p		[85, 87]	CP
Lepilemur edwardsi		6p		[87]	CP
Lepilemur leucopus		1q ter		[87]	CP
Lepilemur microdon		5p		[87]	CP
Lepilemur mittermeieri		7p		[87]	CP
Lepilemur mustelinus		8 ter		[87, 85]	CP
Lepilemur jamesi		5q ter		[87]	CP
Lepilemur ruficaudatus		5q prox		[85, 87]	CP
Lepilemur septentrionalis		14		[85, 87]	CP
Microcebus murinus	Submetacentric	13		[85, 87]	CP
Propithecus verreauxi		6q	5/13	[85]	CP
Otolemur crassicaudatus	Acrocentric	14		[88]	CP
Galago moholi	Metacentric	5	13/16/12	[88]	CP
Otolemur garnettii	Submetacentric	14		[89]	RP
Nycticebus coucang	Submetacentric	18		[89, 90]	RP
		17			CP
Platyrrhini					
Alouatta belzebul	Acrocentric	14		[91]	CP
Alouatta caraya	Acrocentric	15 (20*)		[92, *93]	CP
Alouatta guariba guariba	Acrocentric	14		[93]	CP
Alouatta seniculus arctoidea		16		[91]	CP
Alouatta seniculus macconnelli	Submetacentric	4q	13/19	[92]	CP
Alouatta seniculus sara		12		[91]	CP

	Chromosome type	Chr.	Human association	References	Methods
Aotus lemurinus griseimembra	Acrocentric	17		[93, 94]	CP
Aotus nancymaae	Acrocentric	19		[95]	CP
Ateles geoffroyi				[96]	CP
Ateles belzebuth hybridus	Acrocentric	12		[97]	CP
Ateles belzebuth marginatus	Submetacentric	12		[98]	CP
Ateles paniscus paniscus	Metacentric	4	13a/13b/3c/7b/1a2	[98]	CP
Brachyteles arachnoides	Acrocentric	20		[98]	CP
Callicebus donacophilus pallescens	Acrocentric	15		[99]	CP
Callicebus lugens	Submetacentric	1	1/13–12/13	[100]	CP
Callicebus moloch	Acrocentric	21		[101]	CP
Callicebus cupreus	Submetacentric Acrocentric	7,17	3/21/13, 13/17	[102]	CP
Callimico goeldii	Acrocentrics	19,17	13/9/22, 13/17	[18, 103]	CP RP
Callithrix argentata	Submetacentrics	2,1	13/9/22, 20/17/13	[18, 103]	CP RP
Callithrix jacchus	Submetacentrics	1,5	13/9/22, 20/17/13	[18, 103]	CP RP
Cebuella pygmaea	Submetacentrics	1,4	13/9/22,20/17/13	[18, 103]	CP
Saguinus oedipus	Submetacentrics	1,2	9/13/22,20/17/13	[18, 103]	CP
Cebus apella (Sapajus)	Acrocentric	17		[104, 105]	CP Z-F
Sapajus a. paraguayanus	Acrocentric	17		[105]	Z-F
Sapajus A. robustus	Acrocentric	17		[105]	Z-F
Cebus capucinus	Acrocentric	11		[105]	CP
Cebus nigrivitatus	Acrocentric	17		[97]	CP
Chiropotes israelita	Acrocentric	15		[95]	CP
Chiropotes utahicki	Acrocentric	15		[95]	CP
Lagothrix lagotricha	Submetacentric	8		[106]	CP
Leontopithecus chrysomelas	Submetacentrics	1,2	9/13/22,13/17/20	[107]	CP
Pithecia irrorata	Submetacentric	8	22/13	[108]	CP
Cacajao calvus rubicundus	Acrocentric	13		[108]	CP
Saimiri sciureus	Acrocentric	16		[18, 101]	CP

	Chromosome type	Chr.	Human association	References	Methods
Catarrhini					
Chlorocebus aethiops	Metacentric	3		[109]	CP
Cercopithecus erythrogaster	Submetacentric	12		[110]	Z-F
Cercopithecus neglectus	Metacentric	19		[111]	RP
Cercopithecus stampflii	Submetacentric	13		[110]	Z-F
Presbytis cristata	Metacentric	19		[112]	CP
Colobus guereza	Metacentric	19		[113]	CP
Erythrocebus patas	Submetacentric	15		[111]	RP
Hylobates concolor	Metacentrics	5,9	1/13; 1/4/10/13	[114]	CP
Hylobates klossii		4q	3/13	[115]	CP
Hylobates lar	Metacentric	4q	3/13	[17]	CP
Hylobates moloch		4q	3/13	[115]	CP
Macaca fuscata	Submetacentric	16		[116]	CP
Nasalis larvatus	Metacentric	15		[117]	CP
Pygathrix nemaeus	Submetacentric	17		[118]	CP
Semnopithecus francoisi	Metacentric	9		[119]	CP
Semnopithecus phayrei	Metacentric	9		[115]	CP
Symphalangus syndactylus		15		[17]	CP
Pongo pygmaeus	Acrocentric	14		[17]	CP
Gorilla gorilla	Acrocentric	14		[17]	CP
Pan troglodytes	Acrocentric	14		[17]	CP
Scandentia					
Tupaia belangeri	Acrocentric	17		[120]	CP
Tupaia minor	Acrocentric	16		[121]	CP
Galliformes					
Gallus gallus		1		[51, 52]	SA
Monotremata					
Ornithorhynchus anatinus	Submetacentric Metacentrics	2,10,20		[51]	SA
Didelphimorphia					
Monodelphis domestica	Submetacentrics	4,7		[51, 52]	SA

Table 2. List of species analyzed by chromosomal painting (CP or reciprocal P) and/or sequence alignments (SA) and the references used. For each species is reported the human chromosome 13 homologous and eventually, if present other human associations.

outgroups and data come just from sequence alignments. When HSA 13 synteny, in yellow, is rearranged with just few human syntenies, these are represented in different colours and are reported on the right of the ideogram (e.g., in Indri chromosome 3, synteny 13 is fused

http://dx.doi.org/10.5772/intechopen.79380

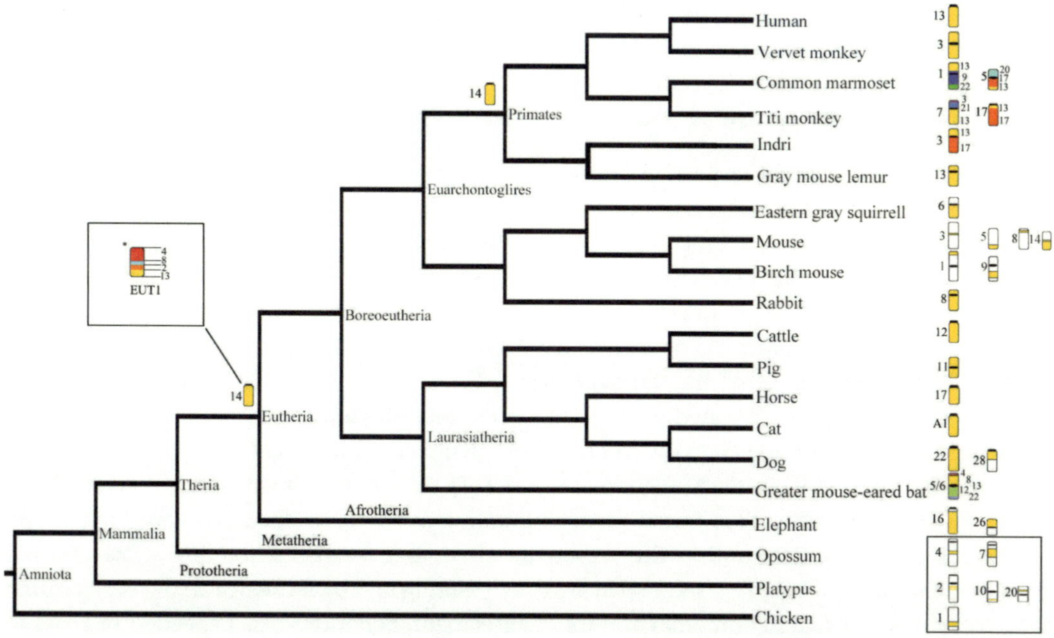

Figure 1. The mammalian phylogenetic tree showing the orthologue blocks that correspond to human chromosome 13—in yellow—in representative eutherian species for which reciprocal chromosome painting is available. For some species also DNA sequence alignments have been previously showed, see **Table 2** for citation; in the tree, it is reported the ancestral synteny 13 form described by painting data analysis and in the box the eutherian ancestral chromosome 13 alternative reconstruction obtained through sequence data* [17]. The platypus (Monotremata), opossum (Metatheria) and chicken (Aves) chromosomes homologues are reported in the box to the low right; these last species are representative outgroups. Different colours represent HSA human syntenies which are reported on the right of the ideogram; white region represents parts of chromosomes covered by many different human syntenies; on the left of the ideogram are reported the species' chromosome number of the 13 human homologues; the black circle is the centromere.

with synteny 17 in red), whereas when HSA 13, in yellow, is rearranged with many human syntenies, these are represented by white segments for logistic issue (e.g., on chicken chromosome 1). Through painting and sequence analysis in mammals, human chromosome 13 has been previously shown to be conserved, with some exceptions (**Table 2**, **Figure 1**). Indeed, the homologues to human chromosome 13 are found as single conserved chromosomes in most representative mammalian orders analysed by chromosome painting, for example in Dermoptera, Pilosa, Carnivori (cat—*Felis silvestris catus* ch A1), Lagomorpha (rabbit—*Oryctolagus cuniculus* ch 8), Perissodactyla (horse—*Equus caballus* ch 17), and Cetartiodactyla (cattle—*Bos tauros* ch 12 and pig—*Sus scrofa* ch 11; in pig, the synteny is metacentric due to a new centromere formation). Human synteny 13 has gone to many rearrangements such as translocation and fission in other different groups; indeed, it is associated with one or more human syntenies due to translocation, as in Tubulidentata, Afrosoricida, Eulipotyphla, Macroscelidea, Sirenia, Pholidota, Chiroptera (**Table 2**). For example, among Chiroptera in Greater mouse-eared bat, on *Myotis myotis* ch 5/6 is present human synteny 13 associated with many other human syntenies (8 lightgreen/4 bordoux/13 yellow/12 green/22 darkgreen) and among Rodentia in eastern grey squirrel, *Sciurus carolinensis* ch 6, human synteny 13 in yellow is associated with other human syntenies (reported in white in **Figure 1**). Furthermore, human synteny 13 is fragmented into two segments or into many segments and associated with other HSA syntenies, for example in Carnivori (Canis—*Canis lupus familiaris* ch 22, 28), in Proboscidea (elephant—*Loxodonta Africana* ch 16, 26), and in Rodentia species such as birch mouse (*Sicista betulina* ch 1, 9; in these last species, many other human syntenic associations are reported in white segments for logistic concern in **Figure 1**).

Through genome assembly analysis (alignments of sequences, SA), chromosome 13 has also been shown to be conserved in many mammals such as pigs, horses, and cats [13], very rearranged in mice (*Mus musculus* ch 3, 5, 8, 14) [14] and fragmented in platypus (*Ornithorhynchus anatinus* ch 2, 10, 20) [51]; moreover, it has also been shown to be present in the outgroups Opossum (*Monodelphis domestica* ch 4, 7) and *chicken* (*Gallus gallus* ch 1) [52] (**Table 2**, **Figure 1**). More recently, researchers analysing more than 19 placental mammals have hypothesised that the eutherian homologue 13 ancestor was fused with other human syntenies (HSA 4, and parts of HSA 2 and 8) [16]. This alternative reconstruction obtained through sequence data (in **Figure 1** reported in the box*) see synteny 13 on EUT ch 1 associated with other HSA syntenies (2 orange/8 lightgreen/4 bordoux) according with previous sequence alignments work [17]. Part of this human associations (13/2/8/4) involving human synteny 13 is found through painting just in Greater mouse-eared bat ch 5/6, HSA syntenies 4/8/13/12/22, and for this reason, the alternative reconstruction do not find support through painting. Thus, the two reconstructions, by painting and by sequence analysis, regarding the ancestral synteny 13 in eutherian are not in agreement. Better analysis is needed in order to clarify this complex origin. The main issue to be considered to shed light on this issue is the use of appropriate outgroups in the reconstruction of the ancestral eutherian chromosome forms and the incomplete set of taxa analysed. Indeed, the lack of comparative chromosome painting between eutherians and other mammals, such as monotreme and marsupials, and on the other hand the lack of data on many genomes do not permit an exact reconstruction [16, 53].

Human chromosome 13 has also been analysed by mapping BAC probes onto representative Mammalian orders [40]; this work has especially focused attention on the history of this chromosome, with particular focus on intrachromosomal rearrangements and the potential relationships between evolutionarily new centromeres (ENCs) and neocentromeres occurring in clinical cases. Indeed, it has been hypothesised that neocentromere formation, a typical event in many tumours, could occur in correspondence to ENC position arising during evolution [54]. BAC mapping has permitted the study of small intrachromosomal rearrangements along the human 13 homologues and the identification of the occurrence of new evolutionary centromeres. Among mammals, evolutionary centromere repositioning on HSA 13 homologues have been shown in pigs and many primates such as for example on *Lagothrix lagotricha* chromosome 8 [40]; furthermore, a small inversion is common in nonprimate mammals [3, 40].

Although human chromosome 13 has been previously shown to be conserved, when compared to ancestral primate chromosomes, it shows some degree of rearrangements in certain primate taxa. Conflicting interpretations of classical banding data on human and great ape chromosome 13 have been published [1, 55, 56]. Among Hominoids, humans, chimpanzees, and orangutans share the same acrocentric form from which the gorillas' differs by only a small paracentric inversion [57]. Among Strepsirrhini, it is a single conserved chromosome as seen for example in grey mouse lemur (*Microcebus murinus* ch 13); however, in this species, synteny 13 is metacentric presumably due to an inversion or alternatively for the occurrence of a new centromere. Synteny 13 has gone to different rearrangements in other species such as, for example, in indri (*Indri indri* ch 3), where it is fused with synteny 17 in red (**Figure 1**).

http://dx.doi.org/10.5772/intechopen.79380

Among Catarrhines (Old World monkeys), the HSA 13 homologues differ in the presence of new centromeres, for example Vervet monkey (*Chlorocebus aethiops*); the *Chlorocebus* chromosome 3 are, indeed, metacentric if compared with the acrocentric human form (**Figure 1**).

Even if human chromosome 13 is presumably conserved in the ancestors of platyrrhines, HSA 13 homologue has gone into many rearrangements in New World monkeys; indeed, synteny 13 has gone to fission and subsequent translocation with other HSA syntenies in Common marmosets (*Callithrix jacchus*), resulting in chromosome 1 and 5 (covered, respectively, by HSA 13 yellow/9 blue/22 darkgreen and 13 yellow/17 red/20 lightgreen), and in Titi monkeys (*Callicebus cupreus*) resulting in ch 7 and 17 (covered, respectively, by HSA 3 fuxia/21 lightblue/13 yellow and 13 yellow/17 red; **Figure 1**). Furthermore, some intrachromosomal rearrangements, such as inversions and new centromeres, have been shown by BAC in other Platyrrhini [40].

3. Conclusion

Classic cytogenetics, using banding, and molecular cytogenetics, using painting or other mapping probes such as BAC, are useful methods for reconstructing human chromosome history in a comparative approach with mammals. Although human chromosome 13 has previously been shown to be conserved in mammals, it is less conserved then previously claimed; indeed, some interchromosomal rearrangements have been demonstrated through painting, and intrachromosomal rearrangements have been shown by BAC mapping in various taxa; for this reason, further analysis is needed. Furthermore, the ancestral eutherian form has yet to be elucidated, as contrasting results continue to be shown through painting and sequence data comparison.

Acknowledgements

Thanks to the "Fondazione Intesa San Paolo Onlus" which supported this work by funding the project "Evoluzione genomica in Primates" (2016-NAZ-0012, CUP, B72F16000130005) of F.D.

Conflict of interest

We have no conflicts of interest.

Author details

Rita Scardino, Vanessa Milioto and Francesca Dumas*

*Address all correspondence to: francesca.dumas@unipa.it

Department of "Scienze e Tecnologie Biologiche, Chimiche e Farmaceutiche (STEBICEF)", University of Palermo, Palermo, Italy

References

[1] Dutrillaux B. Chromosomal evolution in primates: Tentative phylogeny from Microcebus murinus (Prosimian) to man. Human Genetics. 1979;**48**:251-314

[2] Stanyon R, Rocchi M, Capozzi O, Roberto R, Misceo D, Ventura M, et al. Primate chromosome evolution: Ancestral karyotypes, marker order and neocentromeres. Chromosome Research. 2008:17-39

[3] Rocchi M, Archidiacono N, Schempp W, Capozzi O, Stanyon R. Centromere repositioning in mammals. Heredity. 2012;**108**(1):59-67

[4] Dumas F, Mazzoleni S. Neotropical primate evolution and phylogenetic reconstruction using chromosomal data. The Italian Journal of Zoology. 2017;**84**(1):1-18

[5] Sineo L, Dumas F, Vitturi R, Picone B, Privitera O, Stanyon R. Williams-Beuren mapping in *Callithrix argentata, Callicebus cupreus* and *Alouatta caraya* indicates different patterns of chromosomal rearrangements in neotropical primates. Journal of Zoological Systematics and Evolutionary Research. 2007;**45**(4):366-371

[6] Picone B, Dumas F, Stanyon R, Lannino A, Bigoni F, Privitera O, et al. Exploring evolution in Ceboidea (Platyrrhini, primates) by Williams-Beuren probe (HSA 7q11.23) chromosome mapping. Folia Primatologica. 2008;**79**(5):417-427

[7] Dumas F, Sineo L, Ishida T. Taxonomic identification of Aotus (Platyrrhinae) through cytogenetics | Identificazione tassonomica di Aotus (Platyrrhinae) mediante la citogenetica. Journal of Biological Research. 2015;**88**(1):65-66

[8] Dumas F, Cuttaia H, Sineo L. Chromosomal distribution of interstitial telomeric sequences in nine neotropical primates (Platyrrhini): Possible implications in evolution and phylogeny. Journal of Zoological Systematics and Evolutionary Research. 2016; **54**(3):226-236

[9] Mazzoleni S, Schillaci O, Sineo L, Dumas F. Distribution of interstitial telomeric sequences in primates and the pygmy tree shrew (Scandentia). Cytogenetic and Genome Research. 2017;**151**(3):141-150

[10] Mazzoleni S, Rovatsos M, Schillaci O, Dumas F. Cytogenetics evolutionary insight on localization of 18S, 28S rDNA genes on homologous chromosomes in Primates genomes. Comparative Cytogenetics. 2018;**12**(1):27-40

[11] Hruba M, Dvorak P, Weberova L, Subrt I, Hruba M, Dvorak P, et al. Independent coexistence of clones with 13q14 deletion at reciprocal translocation breakpoint and 13q14 interstitial deletion in chronic lymphocytic leukemia. Leukemia & Lymphoma. 2012;**53**(10):2054-2062

[12] Froenicke L. Origins of primate chromosomes—As delineated by Zoo-FISH and alignments of human and mouse draft genome sequences. Cytogenetic and Genome Research. 2005;**108**:122-138

[13] Murphy WJ, Larkin DM, Der WAE, Bourque G, Tesler G, Auvil L, et al. Dynamics of mammalian chromosome evolution inferred from multispecies comparative maps. Science. 2005;**309**(5734):613-617

[14] Ma J, Zhang L, Suh BB, Raney BJ, Burhans RC, Kent WJ, et al. Reconstructing contiguous regions of an ancestral genome. Genome Research. 2006;**16**(12):1557-1565

[15] Robinson TJ, Ruiz-herrera A, Froenicke L. Dissecting the mammalian genome—New insights into chromosomal evolution. Trends in Genetics. 2006;**22**(6):297-301

[16] Kim J, Farré M, Auvil L, Capitanu B, Larkin DM, Ma J, et al. Reconstruction and evolutionary history of eutherian chromosomes. PNAS. 2017;**114**(27):E5379-E5388

[17] Jauch A, Wienberg4 J, Stanyon R, Arnoldt N, Tofanelli S, Ishidaii T, et al. Reconstruction of genomic rearrangements in great apes and gibbons by chromosome painting. PNAS. 1992;**89**:8611-8615

[18] Dumas F, Stanyon R, Sineo L, Stone G, Bigoni F. Phylogenomics of species from four genera of New World monkeys by flow sorting and reciprocal chromosome painting. BMC Evolutionary Biology. 2007;**7**(Suppl 2):S11

[19] Trifonov VA, Stanyon R, Nesterenko AI, Fu B, Perelman PL, O'Brien PCM, et al. Multidirectional cross-species painting illuminates the history of karyotypic evolution in Perissodactyla. Chromosome Research. 2008;**16**(1):89-107

[20] Weise A, Starke H, Mrasek K, Claussen U, Liehr T. New insights into the evolution of chromosome 1. Cytogenetic and Genome Research. 2005;**108**:217-222

[21] Murphy WJ, Froenicke L, O'Brien SJ, Stanyon R. The origin of human chromosome 1 and its homologs in placental mammals. Genome Research. 2003;**13**(8):1880-1888

[22] Fan Y, Linardopoulou E, Friedman C, Williams E, Trask BJ. Genomic structure and evolution of the ancestral chromosome fusion site in 2q13-2q14.1 and paralogous regions on other human chromosomes. Genome Research. 2002;**12**(11):1651-1662

[23] Müller S, Stanyon R, Finelli P, Archidiacono N, Wienberg J. Molecular cytogenetic dissection of human chromosomes 3 and 21 evolution. PNAS. 2000;**97**(1):206-211

[24] Tsend-Ayush E, Grützner F, Yue Y, Grossmann B, Hänsel U, Sudbrak R, et al. Plasticity of human chromosome 3 during primate evolution. Genomics. 2004;**83**(2):193-202

[25] Yue Y, Grossmann B, Tsend-ayush E, Grützner F, Yang F, Haaf T. Genomic structure and paralogous regions of the inversion breakpoint occurring between human chromosome 3p12.3 and orangutan chromosome. Cytogenetic and Genome Research. 2005; **108**(1-3):98-105

[26] Ruiz-Herrera A, Robinson TJ. Evolutionary plasticity and cancer breakpoints in human chromosome 3. BioEssays. 2008;**30**(11-12):1126-1137

[27] Marzella R, Viggiano L, Miolla V, Storlazzi CT, Ricco A, Gentile E, et al. Molecular cytogenetic resources for chromosome 4 and comparative analysis of phylogenetic chromosome IV in great apes. Genomics. 2000;**63**(3):307-313

[28] Dumas F, Sineo L. Chromosomal dynamics in platyrrhinae by mapping BACs probes. Journal of Biological Research. 2012;**LXXXV**:299-301

[29] Dumas F, Sineo L. The evolution of human synteny 4 by mapping sub-chromosomal specific probes in Primates. Caryologia. 2014;**67**(4):281-291

[30] Marzella R, Viggiano L, Ricco AS, Tanzariello A, Fratello A, Archidiacono N, et al. A panel of radiation hybrids and YAC clones specific for human chromosome. Cytogenetics and Cell Genetics. 1997;**77**(3-4):232-237

[31] Szamalek JM, Goidts V, Chuzhanova N, Hameister H, Cooper DN, Keherer-Sawatzki H. Molecular characterization of the pericentric inversion that distinguishes human chromosome 5 from the homologous chimpanzee chromosome. Human Genetics. 2005; **117**(2-3):168-176

[32] Capozzi O, Purgato S, Addabbo PD, Archidiacono N, Battaglia P, Spada F, et al. Evolutionary descent of a human chromosome 6 neocentromere: A jump back to 17 million years ago. Genome Research. 2009;**19**(5):778-784

[33] Eder V, Ventura M, Ianigro M, Teti M, Rocchi M, Archidiacono N. Chromosome 6 phylogeny in primates and centromere repositioning. Molecular Biology and Evolution. 2003;**20**(9):1506-1512

[34] Richard F, Lombard M, Dutrillaux B. Phylogenetic origin of human chromosomes 7, 16, and 19 and their homologs in placental mammals. Genome Research. 2000;**10**(5):644-651

[35] Müller S, Finelli P, Neusser M, Wienberg J. The evolutionary history of human chromosome 7. Genomics. 2004;**84**:458-467

[36] Dumas F, Sineo L. Chromosomal dynamics in Cercopithecini studied by Williams-Beuren probe mapping. Caryologia. 2010;**63**(4):435-442

[37] Montefalcone G, Tempesta S, Rocchi M, Archidiacono N. Centromere repositioning. Genome Research. 1999;**9**(12):1184-1188

[38] Carbone L, Ventura M, Tempesta S. Evolutionary history of chromosome 10 in primates. Chromosoma. 2002:267-272

[39] Cardone MF, Lomiento M, Teti MG, Misceo D, Roberto R, Capozzi O, et al. Evolutionary history of chromosome 11 featuring four distinct centromere repositioning events in Catarrhini. Genomics. 2007;**90**(1):35-43

[40] Cardone MF, Alonso A, Pazienza M, Ventura M, Montemurro G, Carbone L, et al. Independent centromere formation in a capricious, gene-free domain of chromosome 13q21 in Old World monkeys and pigs. Genome Biology. 2006;**7**(10):R91

[41] Ventura M, Mudge JM, Palumbo V, Burn S, Blennow E, Pierluigi M, Giorda R, et al. Neocentromeres in 15q24Y26 map to duplicons which flanked an ancestral centromere in 15q25. Genome Research. 2003;**13**(9):2059-2068

[42] Locke DP, Jiang Z, Pertz LM, Misceo D, Archidiacono N, Eichler EE. Molecular evolution of the human chromosome 15 pericentromeric region. Genome Biology. 2005;**108**:73-82

[43] Misceo D, Ventura M, Eder V, Rocchi M, Archidiacono N. Human chromosome 16 conservation in primates. Chromosome Research. 2003;**11**(4):323-326

[44] Goits V, Szamalek JM, Hameister H, Kehrer-Sawatzki H. Segmental duplication associated with the human-specific inversion of chromosome 18: A further example of the impact of segmental duplications on karyotype and genome evolution in primates. Human Genetics. 2004:116-122

[45] Misceo D, Cardone MF, Carbone L, D'Addabbo P, de Jong PJ, Rocchi M, et al. Evolutionary history of chromosome 20. Molecular Biology and Evolution. 2005;**22**(2):360-366

[46] Wimmer R, Kirsch S, Rappold GA, Schempp W. The evolution of the azoospermia factor region AZFa in higher primates. Cytogenetic and Genome Research. 2005;**108**:211-216

[47] Dunham A, Matthews LH, Burton J, Ashurst JL, Howe KL, Ashcroft KJ, et al. The DNA sequence and analysis of human chromosome 13. Nature. 2004;**428**:522-528

[48] Bailey JA, Gu Z, Clark RA, Reinert K, Samonte RV, Schwartz S, et al. Recent segmental duplications in the human genome. Science. 2002;**297**:1003-1007

[49] Murphy WJ, Eizirik E, Johnson WE, Zhang YP, Ryder OA, O'Brien SJ. Molecular phylogenetics and the origins of placental mammals. Nature. 2001;**409**(6820):614-618

[50] Maddison WP, Maddison DRV. Mesquite: A modular system for evolutionary analysis. 2008;**11**:1103-1118. http.//mesquiteproject

[51] Graphodatsky A, Ferguson-Smith MA, Stanyon R. A short introduction to cytogenetic studies in mammals with reference to the present volume. Cytogenetic and Genome Research. 2012;**137**(2-4):83-96

[52] Robinson TJ, Ruiz-herrera A. Defining the ancestral eutherian karyotype: A cladistic interpretation of chromosome painting and genome sequence assembly data. Chromosome Research. 2008;**16**:1133-1141

[53] Svartman M, Stone G, Page JE, Stanyon R. A chromosome painting test of the basal Eutherian karyotype. Chromosome Research. 2004;**12**:45-53

[54] Alonso A, Mahmood R, Li S, Cheung F, Yoda K, Warburton PE. Genomic microarray analysis reveals distinct locations for the CENP-A binding domains in three human chromosome 13q32 neocentromeres. Human Molecular Genetics. 2003;**12**(20):2711-2721

[55] Clemente IC, Ponsa M, Garcia M, Egozcue J. Evolution of the Simiiformes and the phylogeny of human chromosomes. Human Genetics. 1990;**84**:493-506

[56] Yunis JJ, Prakash O. The origin of man: A chromosomal pictorial legacy. Science. 1982; **215**(4539):1525-1530

[57] Muller S, Wienberg J. "Bar-coding" primate chromosomes: molecular cytogenetic screening for the ancestral hominoid karyotype. 2001;**109**:85-94

[58] Nie W, Fu B, O'Brien PCM, Wang J, Su W, Tanomtong A, et al. Flying lemurs—The "flying tree shrews"? Molecular cytogenetic evidence for a Scandentia-Dermoptera sister clade. BMC Biology. 2008;**6**:18

[59] Yang F, Alkalaeva EZ, Perelman PL, Pardini AT, Harrison WR, O'Brien PCM, et al. Reciprocal chromosome painting among human, aardvark, and elephant (superorder Afrotheria) reveals the likely eutherian ancestral karyotype. PNAS. 2003;**100**:1062-1066

[60] Ruiz-Herrera A, Robinson TJ. Chromosomal instability in Afrotheria: Fragile sites, evolutionary breakpoints and phylogenetic inference from genome sequence assemblies. BMC Evolutionary Biology. 2007;**7**:199

[61] Robinson TJ, Fu B, Ferguson-Smith MA, Yang F. Cross-species chromosome painting in the golden mole and elephant-shrew: Support for the mammalian clades Afrotheria and Afroinsectiphillia but not Afroinsectivora. The Royal Society. 2004;**271**:1477-1484

[62] Kellogg ME, Burkett S, Dennis TR, Stone G, Gray BA, McGuire PM, et al. Chromosome painting in the manatee supports Afrotheria and Paenungulata. BMC Evolutionary Biology. 2007;**7**:6

[63] Ye J, Biltueva L, Huang L, Nie W, Wang J, Jing M, et al. Cross-species chromosome painting unveils cytogenetic signatures for the Eulipotyphla and evidence for the polyphyly of Insectivora. Chromosome Research. 2006;**14**(2):151-159

[64] Yang F, Graphodatsky AS, Li T, Fu B, Dobigny G, Wang J, et al. Comparative genome maps of the pangolin, hedgehog, sloth, anteater and human revealed by cross-species chromosome painting: Further insight into the ancestral karyotype and genome evolution of eutherian mammals. Chromosome Research. 2006;**14**:283-296

[65] Volleth M, Müller S. Zoo-FISH in the European mole (*Talpa europaea*) detects all ancestral Boreo-Eutherian human homologous chromosome associations. Cytogenetic and Genome Research. 2006;**115**:154-157

[66] Svartman M, Stone G, Stanyon R. The ancestral Eutherian karyotype is present in Xenarthra. PLoS Genetics. 2006;**2**(7):1006-1011

[67] Azevedo NF, Svartman M, Manchester A, de Moraes-Barros N, Stanyon R, Vianna-Morgante AM. Chromosome painting in three-toed sloths: A cytogenetic signature and ancestral karyotype for Xenarthra. BMC Evolutionary Biology. 2012;**15-21**:12-36

[68] Cavagna P, Menotti A, Stanyon R. Genomic homology of the domestic ferret with cats and humans. Mammalian Genome. 2000;**11**:866-870

[69] Yang F, O'Brien PCM, Milne BS, Graphodatsky AS, Solanky N, Trifonov V, et al. A complete comparative chromosome map for the dog, red fox, and human and its integration with canine genetic maps. Genomics. 1999;**62**:189-202

[70] Breen M, Thomas R, Binns MM, Carter NP, Langford CF. Reciprocal chromosome painting reveals detailed regions of conserved synteny between the karyotypes of the domestic dog (*Canis familiaris*) and human. Genomics. 1999;**61**(2):145-155

[71] Yang F, Graphodatsky AS, O'Brien PCM, Colabella A, Solanky N, Squire M, et al. Reciprocal chromosome painting illuminates the history of genome evolution of the domestic cat, dog and human. Chromosome Research. 2000;**8**:393-404

[72] Graphodatsky AS, Perelman PL, Sokolovskaya NV, Beklemisheva VR, Serdukova NA, Dobigny G, et al. Phylogenomics of the dog and fox family (Canidae, Carnivora) revealed by chromosome painting. Chromosome Research. 2008;**16**(1):129-143

[73] Perelman PL, Graphodatsky AS, Dragoo JW, Serdyukova NA, Stone G, Cavagna P, et al. Chromosome painting shows that skunks (Mephitidae, Carnivora) have highly rearranged karyotypes. Chromosome Research. 2008;**16**(8):1215-1231

[74] Yang F, Fu B, O'Brien PCM, Nie W, Ryder OA, Ferguson Smith MA. Refined genome-wide comparative map of the domestic horse, donkey and human based on cross-species chromosome painting: Insight into the occasional fertility of mules. Chromosome Research. 2004;**12**:65-76

[75] Nie W, Wang J, Su W, Wang Y, Yang F. Chromosomal rearrangements underlying karyotype differences between Chinese pangolin (*Manis pentadactyla*) and Malayan pangolin (*Manis javanica*) revealed by chromosome painting. Chromosome Research. 2009;**17**:321-329

[76] Balmus G, Trifonov VA, Biltueva LS, O'Brien PCM, Alkalaeva ES, Fu B, et al. Cross species painting among camel, cattle, pig and human: Further insights into the putative Cetartiodactyla ancestral karyotype. Chromosome Research. 2007;**15**:499-514

[77] Kulemzina AI, Trifonov VA, Perelman PL, Rubtsova NV, Volobuev V, Ferguson-Smith MA, et al. Cross-species chromosome painting in Cetartiodactyla: Reconstructing the karyotype evolution in key phylogenetic lineages. Chromosome Research. 2009;**17**(3):419-436

[78] Korstanje R, O'Brien PCM, Yang F, Rens W, Bosma AA, van Lith HA, et al. Complete homology maps of the rabbit (*Oryctolagus cuniculus*) and human by reciprocal chromosomal painting. Cytogenetics and Cell Genetics. 1999;**83**:317-322

[79] Graphodatsky AS, Yang F, Dobigny G, Romanenko SA, Biltueva LS, Perelman PL, et al. Tracking genome organization in rodents by Zoo-FISH. Chromosome Research. 2008;**16**(2):261-274

[80] Stanyon R, Stone G, Garcia M, Froenicke L. Reciprocal chromosome painting shows that squirrels, unlike murid rodents, have a highly conserved genome organization. Genomics. 2003;**82**(2):245-249

[81] Li T, O'Brien PCM, Biltueva L, Fu B, Wang J, Nie W, et al. Evolution of genome organizations of squirrels (Sciuridae) revealed by cross-species chromosome painting. Chromosome Research. 2004;**12**(4):317-335

[82] Volleth M, Heller KG, Pfeiffer RA, Hameister H. A comparative ZOO-FISH analysis in bats elucidates the phylogenetic relationships between Megachiroptera and five microchiropteran families. Chromosome Research. 2002;**10**(6):477-497

[83] Mao X, Nie W, Wang J, Su W, Ao L, Feng Q, et al. Karyotype evolution in *Rhinolophus* bats (Rhinolophidae, Chiroptera) illuminated by cross-species chromosome painting and G-banding comparison. Chromosome Research. 2007;**15**(7):835-847

[84] Mao X, Nie W, Wang J, Su W, Feng Q, Wang Y, et al. Comparative cytogenetics of bats (Chiroptera): The prevalence of Robertsonian translocations limits the power of chromosomal characters in resolving interfamily phylogenetic relationships. Chromosome Research. 2008;**16**(1):155-170

[85] Warter S, Hauwy M, Dutrillaux B, Rumpler Y. Application of molecular cytogenetics for chromosomal evolution of the Lemuriformes (Prosimians). Cytogenetic and Genome Research. 2005;**108**(1-3):197-203

[86] Cardone MF, Ventura M, Tempesta S, Rocchi M, Archidiacono N. Analysis of chromosome conservation in *Lemur catta* studied by chromosome paints and BAC/PAC probes. Chromosoma. 2002;**111**(5):348-356

[87] Rumpler Y, Warter S, Hauwy M, Fausser JL, Roos C, Zinner D. Comparing chromosomal and mitochondrial phylogenies of sportive lemurs (Genus *Lepilemur*, Primates). Chromosome Research. 2008;**16**(8):1143-1158

[88] Stanyon R, Koehler U, Consigliere S. Chromosome painting reveals that galagos have highly derived karyotypes. American Journal of Physical Anthropology. 2002; **117**(4):319-326

[89] Stanyon R, Dumas F, Stone G, Bigoni F. Multidirectional chromosome painting reveals a remarkable syntenic homology between the greater galagos and the slow loris. American Journal of Primatology. 2006;**68**:349-359

[90] Nie W, O'Brien PCM, Fu B, Wang J, Su W, Ferguson-Smith MA, et al. Chromosome painting between human and lorisiform prosimians: Evidence for the HSA 7/16 synteny in the primate ancestral karyotype. American Journal of Physical Anthropology. 2006;**129**(2):250-259

[91] Consigliere S, Stanyon R, Koehler U, Arnold N, Wienberg J. In situ hybridization (FISH) maps chromosomal homologies between *Alouatta belzebul* (Platyrrhini, Cebidae) and other primates and reveals extensive interchromosomal rearrangements between howler monkey genomes. American Journal of Primatology. 1998;**46**(2):119-133

[92] De Oliveira EHC, Neusser M, Figueiredo WB, Nagamachi C, Pieczarka JC, Sbalqueiro IJ, et al. The phylogeny of howler monkeys (*Alouatta*, Platyrrhini): Reconstruction by multicolor cross-species chromosome painting. Chromosome Research. 2002;**10**(8):669-683

[93] Ruiz-Herrera A, Garcia F, Aguilera M, Garcia M, Fontanals MP. Comparative chromosome painting in Aotus reveals a highly derived evolution. American Journal of Primatology. 2005;**65**(1):73-85

[94] Stanyon R, Garofalo F, Steinberg ER, Capozzi O, Di Marco S, Nieves M, et al. Chromosome Painting in two genera of South American monkeys: Species identification, conservation, and management. Cytogenetic and Genome Research. 2011:1-11

[95] Stanyon R, Bigoni F, Slaby T, Müller S, Stone G, Bonvicino CR, et al. Multi-directional chromosome painting maps homologies between species belonging to three genera of New World monkeys and humans. Chromosoma. 2004;**113**(6):305-315

[96] Morescalchi MA, Schempp W, Wienberg J, Stanyon R. Mapping chromosomal homology between humans and the black-handed spider monkey by fluorescence in situ hybridization. Chromosome Research. 1997;**5**(8):527-536

[97] García F, Ruiz-Herrera A, Egozcue J, Ponsà M, Garcia M. Chromosomal homologies between *Cebus* and *Ateles* (Primates) based on ZOO-FISH and g-banding comparisons. American Journal of Primatology. 2002;**57**(4):177-188

[98] De Oliveira EHC, Neusser M, Pieczarka JC, Nagamachi C, Sbalqueiro IJ, Müller S. Phylogenetic inferences of Atelinae (Platyrrhini) based on multi-directional chromosome painting in *Brachyteles arachnoides, Ateles paniscus paniscus* and *Ateles b. marginatus*. Cytogenetic and Genome Research. 2005;**108**(1-3):183-190

[99] Barros RMS, Nagamachi CY, Pieczarka JC, Rodrigues LRR, Neusser M, de Oliveira EH, et al. Chromosomal studies in *Callicebus donacophilus pallescens,* with classic and molecular cytogenetic approaches: Multicolour FISH using human and *Saguinus oedipus* painting probes. Chromosome Research. 2003;**11**(4):327-334

[100] Stanyon R, Bonvicino CR, Svartman M, Seuánez HN. Chromosome painting in *Callicebus lugens,* the species with the diploid number (2n = 16) known in primates. Chromosome Research. 2003;**112**(4):201-206

[101] Stanyon R, Consigliere S, Müller S, Morescalchi A, Neusser M, Wienberg J. Fluorescence in situ hybridization (FISH) maps chromosomal homologies between the dusky titi and squirrel monkey. American Journal of Primatology. 2000;**50**(2):95-107

[102] Dumas F, Bigoni F, Stone G, Sineo L, Stanyon R. Mapping genomic rearrangements in titi monkeys by chromosome flow sorting and multidirectional in-situ hybridization. Chromosome Research. 2005:85-96

[103] Neusser M, Stanyon R, Bigoni F, Wienberg J, Müller S. Molecular cytotaxonomy of New World monkeys (Platyrrhini)—Comparative analysis of five species by multicolor chromosome painting gives evidence for a classification of *Callimico goeldii* within the family of Callitrichidae. Cytogenetics and Cell Genetics. 2001;**94**:206-215

[104] García F, Nogués C, Ponsà M, Ruiz-Herrera A, Egozcue J, Garcia Caldés M. Chromosomal homologies between humans and *Cebus apella* (Primates) revealed by ZOO-FISH. Mammalian Genome. 2000;**11**:399-401

[105] Richard F, Lombard M, Dutrillaux B. ZOO-FISH suggests a complete homology between human and capuchin monkey (Platyrrhini) euchromatin. Chromosome Research. 1996;**36**:417-423

[106] Stanyon R, Consigliere S, Bigoni F, Ferguson-Smith M, O'Brien PCM, Wienberg J. Reciprocal chromosome painting between a New World primate, the woolly monkey, and humans. Chromosome Research. 2001;**9**(2):97-106

[107] Gerbault-Serreau M, Bonnet-Garnier A, Richard F, Dutrillaux B. Chromosome painting comparison of *Leontopithecus chrysomelas* (Callitrichine, Platyrrhini) with man and its phylogenetic position. Chromosome Research. 2004;**12**(7):691-701

[108] Finotelo LFM, Amaral PJS, Pieczarka JC, de Oliveira EHC, Pissinati A, Neusser M, Müller S, Nagamachi CY. Chromosome phylogeny of the subfamily Pitheciinae (Platyrrhini, Primates) by classic cytogenetics and chromosome painting. BMC Evolutionary Biology. 2010;**10**:189

[109] Finelli P, Stanyon R, Plesker R, Ferguson-Smith MA, O'Brien PCM, Wienberg J. Reciprocal chromosome painting shows that the great difference in diploid number between human and African green monkey is mostly due to non-Robertsonian fissions. Mammalian Genome. 1999;**10**(7):713-718

[110] Moulin S, Gerbault-Seureau M, Dutrillaux B, Richard FA. Phylogenomics of African guenons. Chromosome Research. 2008;**16**:783-799

[111] Stanyon R, Bruening R, Stone G, Shearin A, Bigoni F. Reciprocal painting between humans, De Brazza's and patas monkeys reveals a major bifurcation in the Cercopithecini phylogenetic tree. Cytogenetic and Genome Research. 2005;**108**(1-3):175-182

[112] Bigoni F, Koehler U, Stanyon R, Ishida T, Wienberg J. Fluorescence in situ hybridization establishes homology between human and silvered leaf monkey chromosomes, reveals reciprocal translocations bewtween chromosomes homologous to human Y/5, 1/9, and 6/16, and delineates an $X_1X_2Y_1Y_2/X_1X_1X_2X_2$ sex-chromosome system. American Journal of Physical Anthropology. 1997a;**23**:315-327

[113] Bigoni F, Stanyon R, Koehler U, Morescalchi AM, Wienberg J. Mapping homology between human and black and white colobine monkey chromosomes by fluorescent in situ hybridization. American Journal of Primatology. 1997b;**42**(4):289-298

[114] Koehler U, Bigoni F, Wienberg J, Stanyon R. Genomic reorganization in the concolor gibbon (*Hylobates concolor*) revealed by chromosome painting. American Journal of Physical Anthropology. 1995;**292**:287-292

[115] Müller S, Hollatz M, Wienberg J. Chromosomal phylogeny and evolution of gibbons (Hylobatidae). Human Genetics. 2003;**113**(6):493-501

[116] Wienberg J, Stanyon R, Jauch A, Cremer T. Homologies in human and *Macaca fuscata* chromosomes revealed by in situ suppression hybridization with human chromosome specific DNA libraries. Chromosoma. 1992;**101**:265-270

[117] Bigoni F, Stanyon R, Wimmer R, Schempp W. Chromosome painting shows that the proboscis monkey (*Nasalis larvatus*) has a derived karyotype and is phylogenetically nested within Asian colobines. American Journal of Primatology. 2003;**60**(3):85-93

[118] Bigoni F, Houck ML, Ryder OA, Wienberg J, Stanyon R. Chromosome painting shows that *Pygathrix nemaeus* has the most basal karyotype among Asian colobinae. International Journal of Primatology. 2004;**25**(3):679-688

[119] Nie W, Liu R, Chen Y, Wang J, Yang F. Mapping chromosomal homologies between humans and two langurs (*Semnopithecus francoisi* and *S. phayrei*) by chromosome painting. Chromosome Research. 1998;**6**(6):447-453

[120] Müller S, Stanyon R, O'Brien PCM, Ferguson-Smith MA, Plesker R, Wienberg J. Defining the ancestral karyotype of all primates by multidirectional chromosome painting between tree shrews, lemurs and humans. Chromosoma. 1999;**108**(6):393-400

[121] Dumas F, Houck ML, Bigoni F, Perelman P, Romanenko SA, Stanyon R. Chromosome painting of the pygmy tree shrew shows that no derived cytogenetic traits link primates and scandentia. Cytogenetic and Genome Research. 2012;**136**(3):175-179

Chapter 8

Cytogenomic Microarray Testing

Irene Plaza Pinto, Alex da Cruz, Emília Costa,
Samara Pereira, Lysa Minasi and Aparecido da Cruz

Additional information is available at the end of the chapter

http://dx.doi.org/10.5772/intechopen.80514

Abstract

Cytogenomic microarray testing allows the detection of submicroscopic genomic rearrangements, commonly denominated copy number variations (CNVs) that are implicated with many neurodevelopmental disorders, dysmorphic features, multiple congenital anomalies, hematological and solid tumors, and complex disorders and traits in both humans and animals. On the other hand, this approach is also widely used for the identification of structural variations that are applied as a biomarker in pharmacogenomics, agriculture, and animal selection and breeding. The chromosomal microarray analysis (CMA) has been applied for over a decade to screen for submicroscopic genomic gains and losses in DNA sample in both diagnostic and functional scenarios. Herein, we present an overview of the fundamental concepts of cytogenomics and its potential application in human genetic diagnosis, agrigenomics, mutagenesis, and pharmacogenomics.

Keywords: microarray analysis, SNP array, CNV, array CGH

1. Introduction

Since the use of high-resolution chromosome banding and molecular cytogenetic methodologies, several chromosomal aberrations have been identified. Despite genome-wide detection capacity of these technologies, the rearrangements still remain visibly undetectable, which can be explained by microscopic resolution limitations and the lack of knowledge regarding the regions under investigation.

In the 1990s, the array chromosome-based comparative genomic hybridization (array CGH) was established and began to be used for the detection of significant submicroscopic losses and gains with high sensitivity. It was initially applied to analyze copy number changes in tumors, and it was able to detect changes as small as 50 kb in size, with a resolution up to 1000 times higher than the karyotype. Subsequently, the array CGH methodology was optimized and applied to detect unbalanced constitutional rearrangements [1–4]. Initially, bacterial artificial chromosomes (BACs) and fosmid clones were used in array CGH. However, it became clear that not only larger DNA insert constructed in BAC vector but also PCR products and oligonucleotide sequences were good targets for array CGH [5–7]. Microarray approaches offer a high resolution and relatively quick way for genome-wide analysis, increasing the potential possibilities of genomic scrutiny in clinical scenario as well as its potential application in many other distinct structural genome investigations. In addition, chromosomal microarray analysis (CMA) is useful to estimate the breakpoints of the DNA sequences that can reveal potential mechanisms and risk factors underlying the occurrence of chromosome rearrangement, especially in the case of recurrent rearrangements [8]. Nowadays, chromosomal microarray analysis (CMA) is used as a powerful tool to reveal copy number variants thought to play an important role in the pathogenesis of a variety of disorders or the development of complex traits. Thus, genomic variants can be used as a biological biomarker.

Herein, we present an overview of the fundamental concepts of cytogenomics and the potential application of this technology in human genetic diagnosis, agrigenomics, mutagenesis, and pharmacogenomics.

2. Fundamental concepts of cytogenomics: understanding the tool of arrays

Cytogenomic analysis comprises the use of microarray-based technologies for the investigation of specific loci and the entire genome [9]. It has been used for the detection of copy number variation (CNV), defined as genomic intervals that deviate from the normal diploid state that can vary in size ranging from a few base pairs to mega base pairs [10]. The microarray technologies are frequently nominated as chromosomal microarray analysis (CMA), known as comparative genomic hybridization (CGH) and SNP array (**Figure 1**). CMA could be used for a dual role in SNP (single nucleotide polymorphism) and CNV-based association studies and in humans for the evaluation of patients with various diseases and congenital malformations [11, 12].

Initially, the CMA was based on the same principles of chromosome-based comparative genomic hybridization. The CGH was developed in the early 1990s to screen for unbalanced rearrangements in whole genomes [3]. The CGH consists in the simultaneous hybridization of labeled test DNA and normal reference DNA onto normal metaphase chromosomes spread on glass slides. Metaphase CGH was widely used to identify chromosomal numerical alterations associated with solid tumors [13, 14]. Overall, the resolution of metaphase CGH is the same of G-banding cytogenetics, limited to 5–15 Mb. Moreover, experiments require cytogenetic expertise for the preparation of suitable metaphase chromosomes [3]. Due to its resolution limitations, the metaphase CGH became restricted to cancer research and did not demonstrate feasibility for analysis of genomic rearrangements in patients with developmental disorders. Subsequently, CGH was implemented as microarrays replacing the metaphases

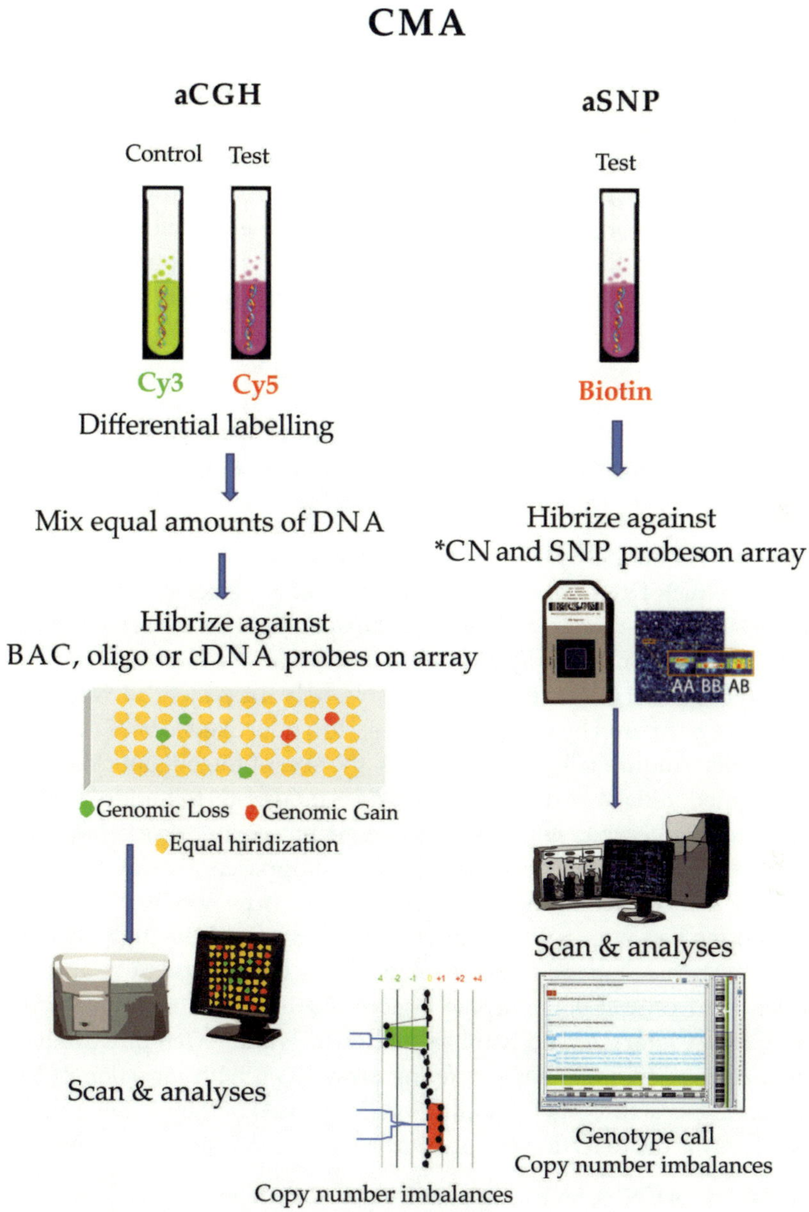

Figure 1. Fundamental concepts of chromosomal microarrays. Currently, the most widely used arrays are comparative genome hybridization arrays (aCGH) and single-nucleotide polymorphism arrays (aSNP). Both microarray-based technologies are able to detect genomic imbalances based on the spectral differences of the fluorescent dyes used to label test and reference DNA in aCGH and test DNA in aSNP. Sample analyses are possible with the aid of computational tools designed to call, view, summarize, and report chromosomal aberrations, including copy number gain or loss, across the genome. *CN: Copy Number.

CGH by cloned DNA segments as genomic reference to be used as targets for the hybridization, improving the potential for the detection of small chromosomal imbalances and increasing its resolution in more than tenfold [14].

The pieces of information regard the DNA sequence of the human genome, which was obtained by the Human Genome Project, making it possible for the construction and characterization of DNA libraries that could be cloned using bacterial artificial chromosomes (BACs) and fosmids as vectors [15, 16]. The array CGH was introduced having known clinically relevant microdeletion genomic DNA segments in BAC clones [10]. However, the relatively large size of the initial clones (170 kb for BACs and 40 kb for fosmids) limited the resolution of the arrays which is dependent on the distances between probes, the length of the clones, and how both probes and clones are distributed across the genome [14, 17]. With the emerging of new protocols, different probes, including small insert clones (1.5–2.5 kb), cDNA clones (0.5–2 kb), PCR products (0.1–1.5 kb), and oligonucleotides (25–85 bp), have been used in the arrays. However, the oligonucleotides have been more appropriated targets for array CGH [5–7, 18]. The oligonucleotide array offers higher resolution and is better than BAC in measuring size of CNVs, increasing the detection of small CNVs. Most CGH arrays available are designed with oligonucleotides ranging from 50 to 70 base pairs (bp) on the probe [19].

According to the International System for Human Cytogenetic Nomenclature [20], in contrast to array CGH, in which DNAs from patient and control are labeled with different dyes and hybridized to a single slide for array CGH and the alterations in the ratio of the two fluorescent dyes indicate a different quantity of DNA in the test sample as compared with the control corresponding locations of CNVs, in the SNP arrays, only the patient's DNA is hybridized to the microarray and compared by computer analysis to a pool of genomic DNA from reference healthy individuals. Additionally, in SNP arrays, the size of the oligonucleotides is about 20 bp and was designed initially to detect genotypes for thousands to hundreds of thousands of SNPs across the entire genome with the focus on genome-wide association studies [19, 21]. SNP arrays can also enable the detection of CNVs, but opposed to array CGH, each probe is located at an SNP and can determine the genotype of the corresponding SNP, and the current SNP arrays with median inter-SNP distances of <0.7 kb ensure the high density of genome coverage [22, 23]. There are many commercial platforms for microarray analysis. Array CGH allows the detection of non-polymorphic region and has been manufactured by Agilent Technologies (Santa Clara, CA) and NimbleGen (Roche Nimble-Gen Inc., Madison, WI), and more recently, Agilent Technologies is offering array CGH with inclusion of SNP markers. The SNP arrays, manufactured by Illumina (San Diego, CA) and by Affymetrix (Santa Clara, CA), have markers for the detection of polymorphic and non-polymorphic regions.

Besides the detection of CNVs, SNP arrays have some advantages in relation to array CGH. The SNP markers can also detect long contiguous stretches of homozygosity (LCSH), low-level mosaic aneuploidies, and chimerism. The detection of LCSH could indicate uniparental isodisomies (UPD) and consanguinity. LCSH distributed in several regions of chromosomes is characterized by genetic identical by descent; on the other hand, when LCSH was identified in a single chromosome, this observation may indicate UPD [24]. The major disadvantage of SNP array and array CGH is the inability to detect balanced chromosome rearrangements because balanced rearrangements show no copy number alterations. Thus, array methodologies do not replace G-banding karyotype for the detection of balanced structural rearrangements. However, they detect abnormalities that are cytogenetically cryptic by G-banded chromosome analysis. It is important to remember that CMA testing cannot detect balanced karyotypic rearrangements such as reciprocal translocations that could be clinically significant if they disrupt a critical gene. For clinical indications with increased risk for a balanced chromosome rearrangement, such as recurrent pregnancy loss, G-banding chromosome analysis should remain a primary diagnostic test [23]. Additionally, in agrigenomics, SNP arrays have facilitated marker-assisted selections, genome-wide association studies, quantitative trait loci analyses, parentage, and traceability, helping in genomic selection programs.

The primary focus for microarrays has been biomedical-related analyses. However, applications for array technology have broadened to include such fields as plant and animal genotyping and pharmacogenomics. The number of CNVs identified has increased as a function of the increased resolution used by the array technologies. The wide use of arrays has allowed their application in agrigenomics providing a powerful and flexible range of genotyping calls useful for genomic selection programs for plants and animals, helping researches and breeders to develop healthier and more productive crops and livestock [25, 26]. Besides, array technologies can be used in pharmacogenomic research for the investigation of potential associations between genomic variation and individual drug response. Several SNP-based microarrays are intended to provide information about specific polymorphisms associated with variable drug responses within individuals in a population, which could increase treatment's overall efficacy and decrease the incidence of adverse events [27, 28]. Moreover, the extensively use of CMA has not only contributed to the identification of CNVs and SNP related to human variability but also contributed to the identification of rearrangements implicated in a variety of diseases such as lifestyle diseases, cancer, autoimmune diseases, and neurodevelopmental disorders, including intellectual disability, autism spectrum disorder, global developmental delay, and neuropsychiatric disorders such as schizophrenia, creating a new field of investigation which has transformed the clinical practice [16, 24, 29].

2.1. Application in the diagnosis of human diseases

Genomic gains and losses, defined as CNVs, often cause a wide variety of specific and complex phenotypes, resulting from alterations in the normal dosage of genes, which cause multiple malformation syndromes, neurodevelopmental disorders (NDD), multiple congenital anomalies (MCA), and dysmorphic features. Nowadays, the improved resolution of the microarray technologies has allowed the identification of cryptic chromosomal alterations, increasing the knowledge of the etiology of genomic disorders and offering potential advantages in the patient's follow-up and management [3, 30].

Since 2010, CMA is widely recognized and recommended as the first-tier cytogenomic diagnostic test for individuals with unexplained developmental delay/intellectual disability (DD/ID), autism spectrum disorders (ASDs), or multiple congenital anomalies, increasing the diagnostic yield around 10–25% [16, 30–32]. DD and ID are defined as several significant delays in developmental areas, including cognitive, speech, social/personal, fine/gross motor, and daily activities. DD is described for children less than 5 years old, and ID is diagnosed at or after the age of 5 years old, with the intelligence quotient less than 70. ASD is a complex spectrum of neurodevelopmental disorders, including autism, Asperger syndrome, pervasive developmental disorder, and childhood disintegrative disorder. MCAs are defined as multiple major structural malformations that cannot be explained by an underlying syndrome or sequence. These disorders might have a genetic etiology involving the gains and losses of CNVs and loss of heterozygosity (LOH), and the clinical consequences of these rearrangements are commonly associated with location, size, and the gene content (**Figures 2** and **3**) [32–34].

In a study with children with ID/DD, ASD, and/or MCAs from Hong Kong, the application of array CGH demonstrated a diagnostic yield of 11% for pathogenic and likely pathogenic CNVs [35]. Another study with a cohort of 339 patients with neurodevelopmental disorders and/or multiple congenital anomalies using the array CGH identified a detection rate of pathogenic CNVs of 20.6% [30]. Combining both array CGH and SNP array in a single platform, it is possible to make the most effective clinical diagnostic offering simultaneously

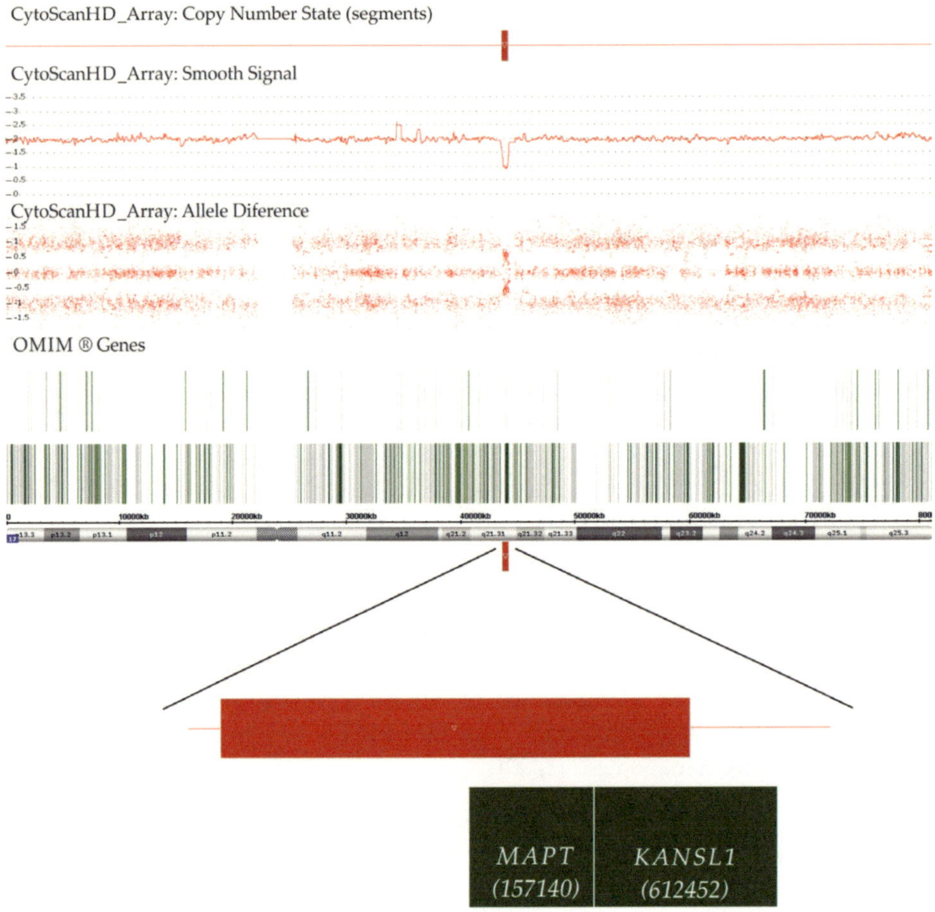

Figure 2. SNP array revealed a copy number loss of chromosome band 17q21.31 of approximately 0.56 Mb in size (arr[hg19] 17q21.31(43.703.801–44.212.416) × 1). This region involves 10 OMIM genes (*LOC644172, CRHR1, MGC57346, C17orf69, MAPT-AS1, SPPL2C, MAPT, MAPT-IT1, STH,* and *KANSL1*) related to Koolen-De Vries Syndrome (MIM610443). The red bar indicates the deleted region and the green bars indicated morbid genes.

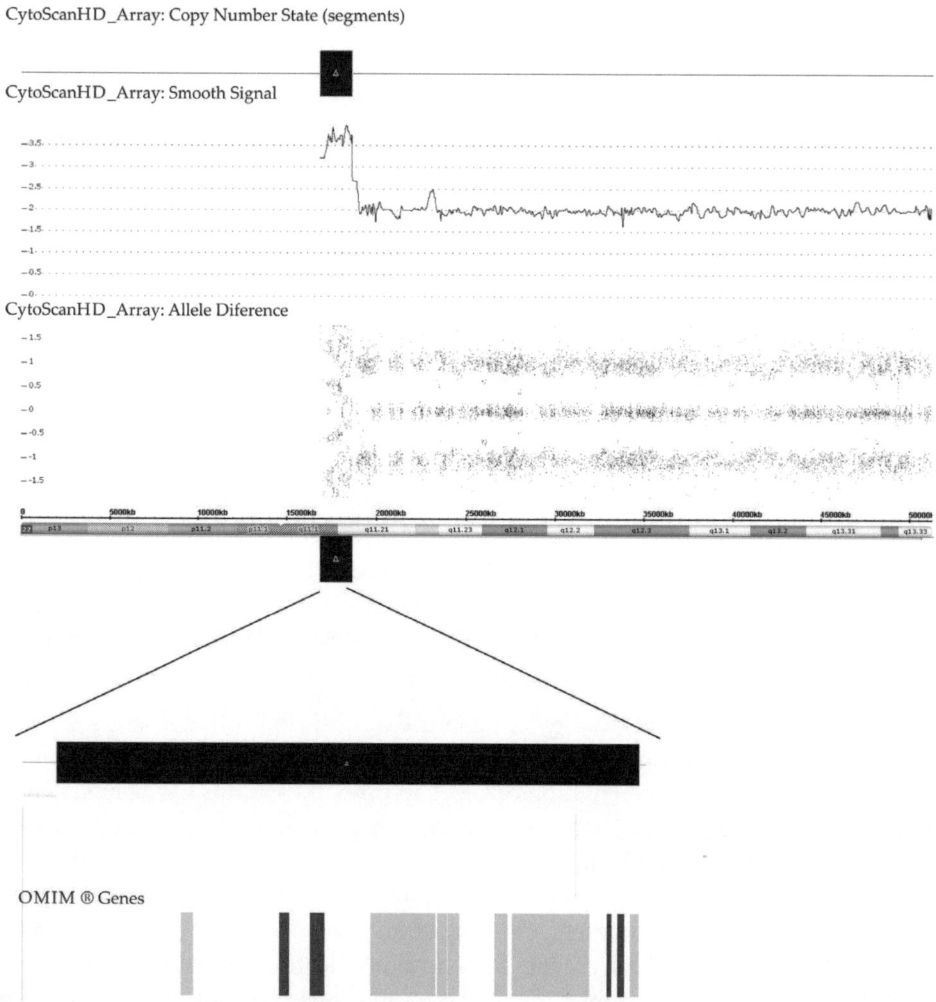

Figure 3. SNP array revealed a gain in copy number of chromosome band 22q11.1, spanning 1.75 Mb (arr[GRCh37] 22q11.1q11.21(16.888.899–18.644.773)×4). This gain as indicated in blue bars and comprise 12 OMIM genes (*XKR3, IL17RA, CECR1, CECR*), *SLC25A18, ATP6V1E1, BID, MICAL3, MIR648, PEX26, TUBA8*, and *USP18*), indicated with gray and green bars, related with Chromosome 22q11.2 Duplication Syndrome (OMIM608363).

identification of CNVs and LOH, as reported by different researches. A comparative study between high-resolution SNP arrays and CGH microarrays revealed that the use of SNP arrays increased the diagnostic yield in children with ID/MCA because these platforms permitted the identification of LOH, which can unravel recessive disorders [36]. Using the combining SNPs with customized exon-targeted oligonucleotide array in a cohort of 3240 patients, Wiszniewska et al. [24] provided a comprehensive approach for the identification of clinically relevant copy number neutral changes in addition to CNVs in a single assay. A study using CMA for 42 Korean patients with unexplained DD, ID, ASD, and MCA identified clinically relevant CNVs in 66.6% of patients [33]. Therefore, microarray-based technologies have become a powerful tool for the identification of genomic rearrangements smaller than 5 Mb that are associated with neurodevelopmental disorders.

In addition to the usefulness of CMA to help increase the diagnostic rate for ID, DD, ASDs, and MCA, a variety of human conditions, such as epilepsy, schizophrenia, bipolar disorder, and attention deficit hyperactivity disorder, have been reported in some individuals whose CNVs have an important causative role [23, 37–42]. Besides the roll of CNVs in the development of neurodevelopmental disorders, CNVs might also contribute to genetic variation in stature in a general population. This conclusion was driven from the observation that in children with short stature lower-frequency copy number variants play a role in the genetic basis of height [43]. Additionally, studies of CNVs in idiopathic short stature (ISS) individuals demonstrated that CMA is a promising approach for the identification of pathogenic CNVs in patients with ISS and could contribute for the recognition of candidate genes associated with growth pathways in humans [44, 45].

CMA testing, especially using SNP-based array, has improved the understanding of genetic causes of many types of human disease and added new pieces of information about gains and losses in the genome in a large variety of hematological malignancies and solid tumors. Additionally, loss of heterozygosity, which is frequently implicated in the tumorigenesis of a variety of cancers, could be detected using SNP array. So, CMA has played an important role to help in diagnosis, prognosis, risk stratification, and therapy for cancer patients [46, 47].

CMA has become a widespread strategy of genetic diagnosis in postnatal settings, especially evaluating children with neurodevelopmental disorders and multiple congenital malformations [48]. Moreover, the implementation of CMA in prenatal settings has helped physicians to identifying chromosomal abnormalities in fetuses harboring anatomical anomalies in the ultrasound, influencing on healthcare delivery in many countries [49]. CMA achieves nearly 100% accuracy rates when applied to identify common aneuploidies in prenatal specimens compared to G-banding karyotyping. Overall, studies showed that in pregnancies with fetal structural anomalies and karyotype with no numerical or structural alterations CMA increased the diagnostic yield around 7%. On the other hand, for all other indications, the increment in the diagnostic yield by CMA has remained around 2% [50, 51]. Taking into consideration the aforementioned information, in 2016 the American Congress of Obstetricians and Gynecologists (ACOG) recommended CMA as the first-tier test for the diagnostic evaluation of fetal structural anomalies. However, the challenge of CMA in prenatal settings is the adequate classification of CNVs as pathogenic and variants of unknown significance (VOUS) [50, 52]. To minimize the reporting of uncertain findings, the practice guideline from Canada issued by the Society of Obstetricians and Gynecologists of Canada (SOGC) and the Canadian College of Medical Geneticists (CCMG) recommended not to report VOUS smaller than 500 kb or VOUS smaller than 1 Mb for losses and gains, respectively [52].

For the diagnosis of human diseases, the microarray platforms should use probes derived from closely spaced genomic loci and have probes concentrated in clinically relevant genes and genomic loci, allowing the detection of smaller CNVs within disease-associated regions. The identification of pathogenic or likely pathogenic CNV by CMA offers benefits for the patient and family, bringing information about prognosis, allowing for appropriate genetic counseling, and adequate patient's management and follow-up for future disclosures [23].

associated with drug responses [79]. Genetic polymorphism studies are performed to classify individuals according to their drug metabolism or disease response capabilities [81].

In a pilot study, Liljedahl et al. [82] developed a microarray genotyping system for multiplex analysis of a panel of SNPs in genes encoding proteins involved in the regulation of blood pressure, demonstrating their viability in response to hypertensive drugs. Therefore, microarray-based tests have provided a useful tool for simultaneous measurement of relative levels of expression of a large number of clinically relevant genes in the context of disease or drug responses [83]. Moreover, the application of the technique in the field of pharmacogenomics characterizes and validates new therapeutic targets, their mechanism of action, metabolic pathways, undesirable side effects, sensitivity, and toxicity to certain drugs [84].

To date, there is a scarcity of studies on the induction of germ line mutations in humans. However, SNP-based arrays can also be applied to monitor individuals exposed to ionizing radiation, and it has been proven to be a useful strategy to evaluate potential health risks related to environmental mutagens. Costa et al. [85] presented results of the analysis of the effect of accidental exposure to low doses of ionizing radiation on the formation of de novo, nonrecurrent CNVs in the progeny of a human population accidentally exposed to cesium-137 during the radiological accident in Goiânia, Brazil. The high-density SNP array used in that study allowed the observation of de novo mutations induced in the germ line of parents exposed to very low doses of ionizing radiation. Although the study of Costa et al. [85] is the pioneer in the field and requires validation, it shed light on the potential of SNP arrays to unravel CVS to be used as useful germ line biomarkers to characterize the exposure of biological systems to mutagenic agents. Thus, a new era of possibilities of using CMA to resolve a variety of biological questions is upon us and once again the future keep on looking promising.

Conflict of interest

The authors declare no conflict of interest.

Author details

Irene Plaza Pinto[1,2], Alex da Cruz[1], Emília Costa[1], Samara Pereira[1], Lysa Minasi[1] and Aparecido da Cruz[1,2,3]*

*Address all correspondence to: acruz@pucgoias.edu.br

1 Pontifical Catholic University of Goiás, School of Agricultural and Biological Sciences, Genetics Master Program, Replicon Research Group, Goiânia, Goiás, Brazil

2 Federal University of Goiás, PhD Program in Biotechnology and Biodiversity, Goiânia, Goiás, Brazil

3 Human Cytogenetics and Molecular Genetics Laboratory, State Laboratory of Public Health Dr. Giovanni Cysneiros, Secretary of Goiás State for Public Health, Goiânia, Goiás, Brazil

References

[1] Albertson DG, Pinkel D. Genomic microarrays in human genetic disease and cancer. Human Molecular Genetics. 2003;**12**:R145-R152. DOI: 10.1093/hmg/ddg261

[2] Cai WW, Mao JH, Chow CW, Damani S, Balmain A, Bradley A. Genome-wide detection of chromosomal imbalances in tumors using BAC microarrays. Nature Biotechnology. 2002;**20**:393-396. DOI: 10.1038/nbt0402-393

[3] Shaffer LG, Bejjani BA. A cytogeneticist's perspective on genomic microarrays. Human Reproduction Update. 2004;**10**:221-226. DOI: 10.1093/humupd/dmh022

[4] Shaw-Smith C, Redon R, Rickman L, Rio M, Willatt L, Fiegler H, et al. Microarray based comparative genomic hybridization (array-CGH) detects submicroscopic chromosomal deletions and duplications in patients with learning disability/mental retardation and dysmorphic features. The Journal of Medical Genetics. 2004;**41**:241-248. PMCID: PMC1735726

[5] Brennan C, Zhang Y, Leo C, Feng B, Cauwels C, Aguirre AJ, Kim M, Protopopov A, Chin L. High resolution global profiling of genomic alterations with long oligonucleotide microarray. Cancer Research. 2004;**64**:4744-4748. DOI: 10.1158/0008-5472.CAN-04-1241

[6] Dhami P, Coffey AJ, Abbs S, Vermeesch JR, Dumanski JP, Woodward KJ, Andrews RM, Langford C, Vetrie D. Exon array CGH: Detection of copy-number changes at the resolution of individual exons in the human genome. American Journal of Human Genetics. 2005;**76**:750-762. DOI: 10.1086/429588

[7] Wolf M, Mousses S, Hautaniemi S, Karhu R, Huusko P, Allinen M, et al. High-resolution analysis of gene copy number alterations in human prostate cancer using CGH on cDNA microarrays: Impact of copy number on gene expression. Neoplasia. 2004;**6**: 240-247. DOI: 10.1593/neo.3439

[8] Weckselblatt B, Rudd MK. Human structural variation: Mechanisms of chromosome rearrangements. Trends in Genetics. 2015;**31**:587-599. DOI: 10.1016/j.tig.2015.05.010

[9] Liehr T, editor. Fluorescence In Situ Hybridization (FISH) - Application Guide. 2nd ed. Jena: Springer; 2017. 588 p

[10] Harel T, Lupski JR. Genomic disorders 20 years on-mechanisms for clinical manifestations. Clinical Genetics. 2018;**93**:439-449. DOI: 10.1111/cge.13146

[11] Haeri M, Gelowani V, Beaudet AL. Chromosomal microarray analysis, or comparative genomic hybridization: A high throughput approach. MethodsX. 2015;**3**:8-18. DOI: 10. 1016/j.mex.2015.11.005

[12] Wang K, Bucan M. Copy Number Variation Detection via High-Density SNP Genotyping. CSH Protoc. 2008; pdb.top46. DOI: 10.1101/pdb.top46

[13] Kallioniemi A, Kallioniemi OP, Sudar DA, Rutovitz D, Gray JW, Waldman F, Pinkel D. Comparative genomic hybridization for molecular cytogenetic analysis of solid tumors. Science. 1992;**258**:818-821. DOI: 10.1126/science.1359641

[14] Redon R, Carter NP. Comparative genomic hybridization: Microarray design and data interpretation. Methods in Molecular Biology. 2009;**529**:37-49. DOI: 10.1007/978-1-59745-538-1_3

[15] Morozova O, Marra MA. From cytogenetics to next-generation sequencing technologies: Advances in the detection of genome rearrangements in tumors. Biochemistry and Cell Biology. 2008;**86**:81-91. DOI: 10.1139/O08-003

[16] Nowakowska B. Clinical interpretation of copy number variants in the human genome. Journal of Applied Genetics. 2017;**58**:449-457. DOI: 10.1007/s13353-017-0407-4

[17] Manning M, Hudgins L, Professional Practice and Guidelines Committee. Array-based technology and recommendations for utilization in medical genetics practice for detection of chromosomal abnormalities. Genetics in Medicine. 2010;**12**:742-745. DOI: 10.1097/GIM.0b013e3181f8baad

[18] Carter NP. Methods and strategies for analyzing copy number variation using DNA microarrays. Nature Genetics. 2007;**39**:S16-S21. DOI: 10.1038/ng2028

[19] Beaudet AL. The utility of chromosomal microarray analysis in developmental and behavioral pediatrics. Child Development. 2013;**84**:121-132. DOI: 10.1111/cdev.12050

[20] McGowan-Jordan J, Simons A, Schmid M, editors. An International System for Human Cytogenomic Nomenclature (ISCN). 1st ed. Switzerland: Karger; 2016. 139 p

[21] Ko JM. Chromosomal microarray: Application for congenital heart diseases. Korean Circulation Journal. 2018;**48**:233-235. DOI: 10.4070/kcj.2018.0032

[22] Keren B. The advantages of SNP arrays over CGH arrays. Molecular Cytogenetics. 2014; **7**(Suppl1):I31. DOI: 10.1186/1755-8166-7-S1-I31

[23] Rosenfeld JA, Patel A. Chromosomal microarrays: Understanding genetics of neurodevelopmental disorders and congenital anomalies. Journal of Pediatric Genetics. 2017; **6**:42-50. DOI: 10.1055/s-0036-1584306

[24] Wiszniewska J, Bi W, Shaw C, Stankiewicz P, Kang SH, Pursley AN, et al. Combined array CGH plus SNP genome analyses in a single assay for optimized clinical testing. The European Journal of Human Genetics. 2014;**22**:79-87. DOI: 10.1038/ejhg.2013.77

[25] Boichard D, Chung H, Dassonneville R, David X, Eggen A, Fritz S, et al. Design of a bovine low-density SNP array optimized for imputation. PLoS One. 2012;**7**:e34130. DOI: 10.1371/journal.pone.0034130

[26] Bovine HapMap Consortium, Gibbs RA, Taylor JF, Van Tassell CP, Barendse W, Eversole KA, et al. Genome-wide survey of SNP variation uncovers the genetic structure of cattle breeds. Science. 2009;**324**:528-532. DOI: 10.1126/science.1167936

[27] Berm EJ, Looff MD, Wilffert B, Boersma C, Annemans L, Vegter S, Boven JF, Postma MJ. Economic evaluations of pharmacogenetic and pharmacogenomic screening tests: A systematic review. Second update of the literature. PLoS One. 2016;**11**:e0146262. DOI: 10.1371/journal.pone.0146262

[28] Matson RS. Applying genomic and proteomic microarray technology in drug discovery. 1st ed. Boca Raton: CRC Press: Taylor & Francis e-Library; 2005. 72 p

[29] Takumi T, Tamada K. CNV biology in neurodevelopmental disorders. Current Opinion in Neurobiology. 2018;**48**:183-192. DOI: 10.1016/j.conb.2017.12.004

[30] Maini I, Ivanovski I, Djuric O, Caraffi SG, Errichiello E, Marinelli M, et al. Prematurity, ventricular septal defect and dysmorphisms are independent predictors of pathogenic copy number variants: A retrospective study on array-CGH results and phenotypical features of 293 children with neurodevelopmental disorders and/or multiple congenital anomalies. Italian Journal of Pediatrics. 2018;**44**:34. DOI: 10.1186/s13052-018-0467-z

[31] Miller DT, Adam MP, Aradhya S, Biesecker LG, Brothman AR, Carter NP, et al. Consensus statement: Chromosomal microarray is a first-tier clinical diagnostic test for individuals with developmental disabilities or congenital anomalies. The American Journal of Human Genetics. 2010;**86**:749-764. DOI: 10.1016/j.ajhg.2010.04.006

[32] Pratte-Santos R, Ribeiro KH, Santos TA, Cintra TS. Analysis of chromosomal abnormalities by CGH-array in patients with dysmorphic and intellectual disability with normal karyotype. Einstein (Sao Paulo). 2016;**14**:30-34. DOI: 10.1590/S1679-45082016AO3592

[33] Lee SH, Song WJ. Chromosomal microarray testing in 42 Korean patients with unexplained developmental delay, intellectual disability, autism spectrum disorders, and multiple congenital anomalies. Genomics and Informatics. 2017;**15**:82-86. DOI: 10.5808/GI.2017.15.3.82

[34] Ooki S. Multiple congenital anomalies after assisted reproductive Technology in Japan (between 2004 and 2009). ISRN Epidemiology. 2013:1-9. DOI: 10.5402/2013/452085

[35] Tao VQ, Chan KY, Chu YW, Mok GT, Tan TY, Yang W, et al. The clinical impact of chromosomal microarray on paediatric care in Hong Kong. PLoS One. 2014;**9**:e109629. DOI: 10.1371/journal.pone.0109629

[36] D'Amours G, Langlois M, Mathonnet G, Fetni R, Nizard S, Srour M, et al. SNP arrays: Comparing diagnostic yields for four platforms in children with developmental delay. BMC Medical Genomics. 2014;**7**:70. DOI: 10.1186/s12920-014-0070-0

[37] Chen J, Calhoun VD, Perrone-Bizzozero NI, Pearlson GD, Sui J, Du Y, Liu J. A pilot study on commonality and specificity of copy number variants in schizophrenia and bipolar disorder. Translational Psychiatry. 2016;**6**:e824. DOI: 10.1038/tp.2016.96

[38] Chen YH, Lu RB, Hung H, Kuo PH. Identifying potential regions of copy number variation for bipolar disorder. Microarrays (Basel). 2014;**3**:52-71. DOI: 10.3390/microarrays3010052

[39] Lowther C, Merico D, Costain G, Waserman J, Boyd K, Noor A, et al. Impact of IQ on the diagnostic yield of chromosomal microarray in a community sample of adults with schizophrenia. Genome Medicine. 2017;**9**:105. DOI: 10.1186/s13073-017-0488-z

[40] Mefford HC. CNVs in epilepsy. Current Genetic Medicine Reports. 2014;**2**:162-167. DOI: 10.1007/s40142-014-0046-6

[41] Thygesen JH, Wolfe K, McQuillin A, Viñas-Jornet M, Baena N, Brison N, et al. Neurodevelopmental risk copy number variants in adults with intellectual disabilities and comorbid psychiatric disorders. The British Journal of Psychiatry. 2018;**212**:287-294. DOI: 10.1192/bjp.2017.65

[42] Vlaskamp DRM, Callenbach PMC, Rump P, Giannini LAA, Dijkhuizen T, Brouwer OF, van Ravenswaaij-Arts CMA. Copy number variation in a hospital-based cohort of children with epilepsy. Epilepsia Open. 2017;**2**:244-254. DOI: 10.1002/epi4.12057

[43] Dauber A, Yu Y, Turchin MC, Chiang CW, Meng YA, Demerath EW, et al. Genome-wide association of copy-number variation reveals an association between short stature and the presence of low-frequency genomic deletions. American Journal of Human Genetics. 2011;**89**:751-759. DOI: 10.1016/j.ajhg.2011.10.014

[44] Hu G, Fan Y, Wang L, Yao RE, Huang X, Shen Y, Yu Y, Gu X. Copy number variations in 119 Chinese children with idiopathic short stature identified by the custom genome-wide microarray. Molecular Cytogenetics. 2016;**9**:16. DOI: 10.1186/s13039-016-0225-0

[45] Singh H, Tiwari P, Bhavi V, Chaudhary PS, Suravajhala P, Mohan MK, Mathur SK. Application of chromosomal microarray for evaluation of idiopathic short stature in Asian Indian children: A pilot study. Indian Journal of Endocrinology and Metabolism. 2018;**22**:100-106. DOI: 10.4103/ijem.IJEM_202_17

[46] Mukherjee S, Sathanoori M, Ma Z, Andreatta M, Lennon PA, Wheeler SR, et al. Addition of chromosomal microarray and next generation sequencing to FISH and classical cytogenetics enhances genomic profiling of myeloid malignancies. Cancer Genetics. 2017;**216-217**:128-141. DOI: 10.1016/j.cancergen.2017.07.010

[47] Nowak D, Hofmann WK, Koeffler HP. Genome-wide mapping of copy number variations using SNP arrays. Transfusion Medicine and Hemotherapy. 2009;**36**:246-251. DOI: 10.1159/000225372

[48] de Jong A, Dondorp WJ, Macville MV, de Die-Smulders CE, van Lith JM, de Wert GM. Microarrays as a diagnostic tool in prenatal screening strategies: Ethical reflection. Human Genetics. 2014;**133**:163-172. DOI: 10.1007/s00439-013-1365-5

[49] Callaway JL, Shaffer LG, Chitty LS, Rosenfeld JA, Crolla JA. The clinical utility of microarray technologies applied to prenatal cytogenetics in the presence of a normal conventional karyotype: A review of the literature. Prenatal Diagnosis. 2013;**33**:1119-1123. DOI: 10.1002/pd.4209

[50] Levy B, Wapner R. Prenatal diagnosis by chromosomal microarray analysis. Fertility and Sterility. 2018;**109**:201-212. DOI: 10.1016/j.fertnstert.2018.01.005

[51] Wapner RJ, Martin CL, Levy B, Ballif BC, Eng CM, Zachary JM, et al. Chromosomal microarray versus karyotyping for prenatal diagnosis. The New England Journal of Medicine. 2012;**367**:2175-2184. DOI: 10.1056/NEJMoa1203382

[52] Armour CM, Dougan SD, Brock JA, Chari R, Chodirker BN, DeBie I, et al. Practice guideline: Joint CCMG-SOGC recommendations for the use of chromosomal microarray

analysis for prenatal diagnosis and assessment of fetal loss in Canada. Journal of Medical Genetics. 2018;**55**:215-221. DOI: 10.1136/jmedgenet-2017-105013

[53] Bhat SA, Malik AA, Ahmad SM, Shah RA, Ganai NA, Shafi SS, Shabir N. Advances in genome editing for improved animal breeding: A review. Veterinary World. 2017;**10**: 1361-1366. DOI: 10.14202/vetworld.2017.1361-1366

[54] Gonen S, Jenko J, Gorjanc G, Mileham AJ, Whitelaw CB, Hickey JM. Potential of gene drives with genome editing to increase genetic gain in livestock breeding programs. Genetics, Selection, Evolution. 2017;**49**:3. DOI: 10.1186/s12711-016-0280-3

[55] Carroll D, Charo RA. The societal opportunities and challenges of genome editing. Genome Biology. 2015;**16**:242. DOI: 10.1186/s13059-015-0812-0

[56] Blendon S, Gorski M, Benson J. The public and the gene-editing revolution. The New England Journal of Medicine. 2016;**374**:1406-1411. DOI: 10.1056/NEJMp1602010

[57] Murray JD, Maga EA. Genetically engineered livestock for agriculture: A generation after the first transgenic animal research conference. Transgenic Research. 2016;**25**: 321-327. DOI: 10.1007/s11248-016-9927-7

[58] Fleming A, Emhimad AA, Christian M, Christine FB. Invited review: Reproductive and genomic technologies to optimize breeding strategies for genetic progress in dairy cattle. Archives Animal Breeding. 2018;**61**:43-57. DOI: 10.5194/aab-61-43-2018

[59] Davoudi P, Abdollahi-Arpanahi R, Nejati-Javaremi A. The impact of QTL allele frequency distribution on the accuracy of genomic prediction. Archives Animal Breeding. 2018;**61**:207-213. DOI: DOI.org/10.5194/aab-61-207-2018

[60] de Los Campos G, Vazquez AI, Fernando R, Klimentidis YC, Sorensen D. Prediction of complex human traits using the genomic best linear unbiased predictor. PLoS Genetics. 2013;**9**:e1003608. DOI: 10.1371/journal.pgen.1003608

[61] Schaefer RJ, Schubert M, Bailey E, Bannasch DL, Barrey E, Bar-Gal GK, et al. Developing a 670k genotyping array to tag ~2M SNPs across 24 horse breeds. BMC Genomics. 2017;**18**:565. DOI: 10.1186/s12864-017-3943-8

[62] Júnior GAO, Perez BC, Ferraz JBS. Genomics applied to puberty in beef cattle (Bos Indicus). Revista Brasileira de Reprodução Animal. 2017;**41**:264-269

[63] Gutiérrez-Gil B, Arranz JJ, Pong-Wong R, García-Gámez E, Kijas J, Wiener P. Application of selection mapping to identify genomic regions associated with dairy production in sheep. PLoS One. 2014;**9**:e94623. DOI: 10.1371/journal.pone.0094623

[64] Kemper KE, Goddard ME. Understanding and predicting complex traits: knowledge from cattle. Human Molecular Genetics. 2012;**21**(R1):R45-51. DOI: 10.1093/hmg/dds332

[65] Bruford MW, Ginja C, Hoffmann I, Joost S, Orozco-terWengel P, Alberto FJ, et al. X. Prospects and challenges for the conservation of farm animal genomic resources, 2015-2025. Frontiers in Genetics. 2015;**21**:314. DOI: 10.3389/fgene.2015.00314

[66] da Cruz, Alex Silva. Estudo de associação ampla do genoma bovino para lactação ajustada em 305 dias em Girolando [thesis]. Goiânia: Universidade Federal de Goiás; 2015. p. 106

[67] Bosch P, Forcato DO, Alustiza FE, Alessio AP, Fili AE, Olmos Nicotra MF, Liaudat AC, Rodríguez N, Talluri TR, Kues WA. Exogenous enzymes upgrade transgenesis and genetic engineering of farm animals. Cellular and Molecular Life Sciences. 2015;**72**: 1907-1929. DOI: 10.1007/s00018-015-1842-1

[68] Mrode R, Tarekegn GM, Mwacharo JM, Djikeng A. Invited review: Genomic selection for small ruminants in developed countries: How applicable for the rest of the world? Animal. 2018;**12**:1333-1340. DOI: 10.1017/S1751731117003688

[69] Uemoto Y, Sasaki S, Kojima T, Sugimoto Y, Watanabe T. Impact of QTL minor allele frequency on genomic evaluation using real genotype data and simulated phenotypes in Japanese black cattle. BMC Genetics. 2015;**16**:134. DOI: 10.1186/s12863-015-0287-8

[70] Hickey JM. Sequencing millions of animals for genomic selection 2.0. Journal of Animal Breeding and Genetics. 2013;**130**:331-332. DOI: 10.1111/jbg.12054

[71] Auvray B, McEwan J, Newman SA, Lee M, Dodds K. Genomic prediction of breeding values in the New Zealand sheep industry using a 50K SNP chip. Journal of Animal Science. 2014;**92**:4375-4389. DOI: 10.2527/jas.2014-7801

[72] Teng X, Xiao H. Perspectives of DNA microarray and next-generation DNA sequencing technologies. Science in China. Series C, Life Sciences. 2009;**52**:7-16. DOI: 10.1007/s11427-009-0012-9

[73] Steyaert W, Callens S, Coucke P, Dermaut B, Hemelsoet D, Terryn W, Poppe B. Future perspectives of genome-scale sequencing. Acta Clinica Belgica. 2018;**73**:7-10. DOI: 10.1080/17843286.2017.1413809

[74] Aardema MJ, MacGregor JT. Toxicology and genetic toxicology in the new era of "toxicogenomics": Impact of "-omics" technologies. Mutation Research. 2002;**499**:13-25. DOI: 10.1016/S0027-5107(01)00292-5

[75] Guindalini C, Tufik S. Use of microarrays in the search of gene expression patterns - application to the study of complex phenotypes. Revista Brasileira de Psiquiatria. 2007;**29**:370-374. DOI: 10.1590/S1516-44462007000400014

[76] Saei AA, Omidi Y. A glance at DNA microarray technology and applications. BioImpacts: BI. 2011;**1**:75-86. DOI: 10.5681/bi.2011.011

[77] Chavan P, Joshi K, Patwardhan B. DNA microarrays in herbal drug research. Evidence-based Complementary and Alternative Medicine. 2006;**3**:447-457. DOI: 10.1093/ecam/nel075

[78] Bumgarner R. DNA microarrays: Types, applications and their future. Current Protocols in Molecular Biology. 2013;**22**:1-17. DOI: 10.1002/0471142727.mb2201s101

[79] Ventola CL. Role of Pharmacogenomic biomarkers in predicting and improving drug response part 1: The clinical significance of pharmacogenetic variants. P&T. 2013;**38**: 545-551. PMID: 24273401

[80] Wang Z, Du Q, Wang F, Liu Z, Li B, Wang A, Wang Y. Microarray analysis of gene expression on herbal glycoside recipes improving deficient ability of spatial learning memory in ischemic mice. Journal of Neurochemistry. 2004;**88**:1406-1415. DOI: 10.1046/j.1471-4159.2003.02258.x

[81] Shahandeh A, Johnstone DM, Atkins JR, Sontag JM, Heidari M, Daneshi N, Freeman-Acquah E, Milward EA. Advantages of array-based technologies for pre-emptive pharmacogenomics testing. Microarrays. 2016;**5**:1-11. DOI: 10.3390

[82] Liljedahl U, Karlsson J, Melhus H, Kurland L, Lindersson M, Kahan T, Nyström F, Lind L, Syvänen AC. A microarray minisequencing system for pharmacogenetic profiling of antihypertensive drug response. Pharmacogenetics. 2003;**3**:7-17. PMID: 12544508

[83] Anderson DC, Kodukula K. Biomarkers in pharmacology and drug discovery. Biochemical Pharmacology. 2014;**87**:172-188. DOI: 10.1016/j.bcp.2013.08.026

[84] Meloni R, Khalfallah O, Biguet NF. DNA microarrays and pharmacogenomics. Pharmacological Research. 2004;**49**:303-308. DOI: 10.1016/j.phrs.2003.06.001

[85] Costa EOA, Pinto IP, Gonçalves MW, da Silva JF, Oliveira LG, da Cruz AS, et al. Small de novo CNVs as biomarkers of parental exposure to low doses of ionizing radiation of caesium-137. Scientific Reports. 2018;**8**:1-13. DOI: 10.1038/s41598-018-23813-5